I0044337

Introductory Geomorphology

Introductory Geomorphology

Greg Powell

Introductory Geomorphology
Greg Powell
ISBN: 978-1-64172-098-4 (Hardback)

Published by Larsen and Keller Education,
5 Penn Plaza,
19th Floor,
New York, NY 10001, USA

Cataloging-in-Publication Data

Introductory geomorphology / Greg Powell.
p. cm.
Includes bibliographical references and index.
ISBN 978-1-64172-098-4
1. Geomorphology. 2. Physical geography. I. Powell, Greg.
GB401.5 .I58 2019
551.41--dc23

Table of Contents

Preface

Geomorphology studies the various physical, biological and chemical processes that act on the Earth's surface to create a variety of topographic and bathymetric features. The study of the origin and evolution of these features is also under the scope of this field. Such studies are aided by physical experiments, field observations and numerical modeling. It is an integrated study of the disciplines of geodesy, geology, engineering geology, geotechnical engineering, physical geography and archaeology. The primary factors responsible for the creation of topographic features of the earth are the wind, surface water flow, chemical dissolution, tectonism, volcanism and glacial action. Some other processes affecting them include periglacial processes, marine currents, extraterrestrial impact, salt-mediated action and seepage of fluids through the seafloor. This textbook provides comprehensive insights into the field of geomorphology. It presents this complex subject in the most comprehensible and easy to understand language. For all those who are interested in geomorphology, this book can prove to be an essential guide.

Given below is the chapter wise description of the book:

Chapter 1, The Earth's surface is characterized by various bathymetric and topographic features, created due to chemical, biological and physical processes operating at or near the surface. The study of the origin and evolution of these features and a prediction of the changes to landforms is under the scope of geomorphology. This chapter has been carefully written to provide an introduction to geomorphology and discusses the varied aspects like biogeomorphology, climatic geomorphology, hydrogeomorphology, applied geomorphology, etc. **Chapter 2**, Geomorphological processes are responsible for the creation and change to the topographic features. Some of these are tectonism, volcanism, groundwater movement, surface water flow, air movement, glacial action, etc. Such processes can be categorized under different groups such as aeolian, biological, fluvial, glacial and igneous processes, among others. The topics elucidated in this chapter cover some of these important processes for an in-depth understanding of geomorphology. **Chapter 3**, Landforms are naturally-occurring features of the Earth. These are categorized according to the factors of elevation, slope, orientation, rock exposure, stratification, etc. This chapter closely examines some of the crucial geomorphological landforms, such as Aeolian, glacial, fluvial, coastal, volcanic and slope landforms. **Chapter 4**, Weathering refers to the slow breaking down of rocks, minerals, soil, wood, etc. due to to their contact with the Earth's water, atmosphere and organisms. There are two important classifications of weathering, physical and chemical weathering. Each of these can involve a biological influence. An elaborate study of the varied processes responsible for weathering, such as physical, biological and chemical weathering, has been explained in this chapter. **Chapter 5**, The action of surface processes that transports dissolved material, soil or rock from one location to another is known as erosion. Deposition is another geological process that adds sediments, rocks and soil to a land mass. It is caused due to wind, water, ice or gravity transport. This chapter explores the different types of erosion and deposition processes such as coastal, internal, soil, sheet, glacial erosion, etc.

At the end, I would like to thank all those who dedicated their time and efforts for the successful completion of this book. I also wish to convey my gratitude towards my friends and family who supported me at every step.

Greg Powell

1

Geomorphology and its Branches

The Earth's surface is characterized by various bathymetric and topographic features, created due to chemical, biological and physical processes operating at or near the surface. The study of the origin and evolution of these features and a prediction of the changes to landforms is under the scope of geomorphology. This chapter has been carefully written to provide an introduction to geomorphology and discusses the varied aspects like biogeomorphology, climatic geomorphology, hydrogeomorphology, applied geomorphology, etc.

Geomorphology is the scientific study of the surface of a planet and those processes responsible for forming it. Scientists involved in this field often study historical changes, through events such as erosion, in order to understand how a particular geographical region came into existence. They may also study current data to better predict how landforms might change in the future and to understand how people can help maintain current features. This allows scientists to anticipate changes in the general structure of the earth.

Natural Processes of Change

Landforms on any world, including earth, are not static; they are part of a dynamically changing system. There are various geomorphic processes that can alter the surface of a world, including plate tectonics, changes in climate, and human activities. Wind can shape landscapes, as can water — both liquid and ice, in the form of glaciers. Volcanic activity, including violent eruptions and the steady flow of lava from some sites, can create new islands or devastate a landscape. Plants and animals can also alter landforms, whether a beaver damming a river or a grove of trees that anchor the soil in a particular location.

Tectonic Changes

Slow movements of the earth's tectonic plates contribute to the uplift and elevation of landforms. There are two common types of tectonic uplift: orogenic and isostatic. Orogenic tectonic uplift is caused when tectonic plates crash together, which raises the land where they meet to create forms such as mountains. Isostatic uplift, on the other hand, refers to how landforms can become higher after the weight on the land is reduced; as land is eroded or glaciers melt, it is believed that the land that was being weighed down can rise.

Changes Caused by Water

The geomorphic effects of water bodies are studied in fluvial geomorphology, which examines how bodies of water alter the landscape. As waterways such as rivers flow, they often carry sediment, which reduces the land around the river itself but increases areas where this sediment is released. Water from rain and flash floods can also be responsible for erosion, which physically alters rocks and other land areas.

Changes Caused by Glaciers

Glaciers also change the landscape. As these heavy sheets of ice advanced across the landscape during the last ice age, they scoured the softer land areas in their way; they also picked up some of this material and moved it. When the ice melted, valleys and fjords coastal valleys that are filled with water were left behind, as were the rocks and soil, called “till,” that the glacier picked up.

Volcanic Changes

On the opposite side, volcanoes can both create and destroy landforms. Often found at the edges of tectonic plates, underwater volcanoes have shaped islands like Hawaii, the Philippine Islands, and New Zealand. On land, they can form large volcanic mountains. The violent explosion of a volcano can radically change the landscape, and wipe out plants and animals in the area.

Changes Caused by Wind

Although it often works much more slowly, wind can also alter the land. Called eolian geomorphology, wind can erode landforms, breaking them down, and build others up, as material is moved from one place to another. The Nebraska Sand Hills, for example, is an area where ancient winds created huge sand dunes that have since stabilized and became a regular part of the landscape.

Biogeomorphology

Plants and animals can have a big impact on the landscape as well. Animals dig tunnels and dens, move rocks and soil, and block rivers, among other things. Plant roots can grow through the cracks in rocks, breaking them apart, or help to hold the soil in an area together, decreasing erosion caused by water and wind. Living things can also combine with other forces to cause changes; a volcanic eruption may destroy a stand of trees, for example, leaving the land in the area exposed to the weathering caused by wind and rain.

Changes Caused by People

Human interventions can also contribute to changes on the earth. With the expansion of civilization, humans began to enact direct changes to their surroundings. The most radical changes to landforms are possible due to technological and organizational advances; the building of the Panama or Suez canals, for example, were significant alterations to the earth’s natural form. People have straightened rivers or prevented them from naturally changing their course, created lakes and other bodies of water, and prevented beaches from expanding or eroding in some cases. The long-term effects of many of the changes that human beings have made is not fully known, and it may take centuries for the side effects good and bad to become fully clear.

Geomorphology on Other Planets

Geomorphology is not restricted to questions about landforms on earth; it applies to all terrestrial planets. The field of “extraterrestrial geomorphology” is expanding due to the influx of scientific data from satellites and space expeditions. For example, volcanic eruptions on the surface of Io, one of Jupiter’s moons, have created many unique features, including tall mountains and flat plains. Scientists studying Mars, Venus, and other planets examine formations like channels and valleys, and theorize about the different processes that might have created them.

Related Fields of Study

A number of different research methods and fields of inquiry are often used for this type of study. Archaeology, for example, can be invaluable in understanding how past human populations have changed and shaped the environment and geography. Study of global surveys through ground and orbital photography is also beneficial, as it allows geomorphologists to gain a better perspective of various landforms. Soil scientists investigate the composition and formation of soils, which helps explain how an area has changed over time.

Geomorphology is also used in a number of other fields. Civil engineers, for example, build and maintain structures like roads, bridges, and dams; understanding how the landscape was formed and how it may change is crucial for such projects. Environmental resource management involves finding ways to make the best use of resources, including water and land, so an understanding of how human activities can change those resources is vital.

Geomorphic system is a set of related landforms and processes, usually defined in terms of a dominant agent of geomorphic activity (water, gravity, ice, wind, waves, or organisms).

Geomorphic Process

Geomorphic Process is the dominant internal or external geologic force that has interacted with the existing geologic structural framework to shape the Earth's surface. Geomorphic Process has two hierarchical elements within this classification:

Geomorphic process type and geomorphic subprocess. In addition, a subprocess modifier is used as a further subdivision for some geomorphic subprocess types. The definitions for each are as follows.

- Geomorphic process type - a general description of the dominant geomorphic process responsible for the nature, origin and development of the landforms. Geomorphic process types are fluvial, glacial, periglacial, lacustrine, tectonic, volcanic, mass wasting, coastal marine, solution, and eolian.
- Geomorphic subprocess - a subdivision of geomorphic process which groups related landforms. For example, the glacial geomorphic process type is subdivided into ice erosion, melt water erosion, water deposition (in close proximity to ice), ice deposition, active ice and snow features, and proglacial deposition.
- Subprocess modifier - a subdivision of subprocess used for the fluvial and glacial process types. For example, alpine and continental are subprocess modifiers for each of the glacial subprocesses.

The effects of water are fluvial processes while the effects of moving tectonic plates are tectonic processes. These, along with the processes listed below, are just some of the systems geomorphologists use to understand the Earth's shape:

- Igneous processes: Volcanoes and the molten rock below the Earth's surface help to form its shape. It's fairly obvious how the molten rock produced by a volcanic eruption can change the shape of the Earth, but other features are formed by magma slowly coming out of cracks in the crust, and even magma moving under the surface can make land move up and down.
- Hill slope processes: It might surprise you, but some people really do study how rocks roll down hills. The way earth and rock moves down slopes can make a big difference to the Earth's shape: in particular, if a hill gets too steep a lot of material can come down very quickly, restricting how quickly it can grow.
- Aeolian processes: Even the wind affects geomorphology. This can be seen easily in deserts or at some beaches, where sand dunes shift around in the wind, but it happens everywhere. For example, where there are large, soft sandstones, the wind can erode large depressions called alcoves.
- Biological processes: Have you ever seen an old tree blown over in a storm? They can leave behind a huge hole where their roots have been ripped up. This is called "tree throw", and it's one example of a biological geomorphic process. Biological processes range from animal burrows to the way living things affect the climate which, in turn, affects the landscape.
- Human processes: Whether it's building canals, carving faces in Mount Rushmore or blowing holes in the ground with bombs, humans change the landscape in all sorts of ways, which geomorphologists take into account.
- Extraterrestrial processes: No, it's not aliens writing interplanetary graffiti on the Earth – although that would be part of geomorphology too! If you've ever seen a picture of a meteorite impact crater, you'll know that objects from outer space can make a big difference to the shape of the planet. One crater in South Africa is 300km across and 2020 million years old.

Landform

A landform is defined as "Any physical feature of the earth's surface having a characteristic, recognizable shape and produced by natural causes". The landform component of the geomorphic classification is directly linked to the geomorphic process element described above in a hierarchical manner. The landform portion of the classification consists of two hierarchical components: landform and element landform. Following are definitions of each.

Landform - a landform that exists within a single geomorphic process type, and which can be delineated at scales at or above the land unit (land type and land type phase) level. For instance, under the subprocess of proglacial depositional, landforms include outwash plain, outwash fan, outwash terrace kettled outwash plain, valley train and outburst floodplain.

Element landform - a spatial component of a landform at the level immediately below, and hierarchical to landform, which can be delineated at scales at or above the land unit (land type and land type phase) level. A beach ridge is a subdivision of a beach and a toe zone is a subdivision of an earth flow. Table is an example of element landforms associated with a glacial trough landform.

Geomorphic Process	Subprocess	Subprocess Modifier	Landform	Element Landforms
	Active Ice and Snow Processes	Alpine	Glacier	Moulin Serac Pressure Ridge Crevasse Bergshrund Ice Apron Drainage Channel (Undifferenctiated) Fosse

Table: An example of the hierarchical link between Geomorphic Process and Landform.

Not all landforms have element landforms due either to the relative simplicity of the classification system adopted for that particular geomorphic process type, or the nature of the geomorphic process type.

The following classification constituents are not hierarchically linked to the classification, but are important components of the classification system:

Landscape term - a general or descriptive term used to describe collections of landforms, generated from a variety of geomorphic processes, on a sub regional or landscape basis, e.G., Basin and range, bolson, glaciated uplands, valleys, canyonland, plateau, and ridge and valley.

Common landform - landforms that commonly occur as components of a variety of larger landforms. These are landforms not clearly tied to a single geomorphic process type or landform. For instance, the common landform swale may occur on moraines, toe zones, eroding hill slopes, or alluvial fans. They represent terms that are "common usage" or "generic" such as mound, knob, or bench. Common landforms should be used in conjunction with a process landform unless they are used to describe a site (point or plot) where the process and landform cannot be determined.

Microfeature - small, local forms on the land surface with an aerial extent (for individual features) of less than a few meters, that can be described by vertical changes in elevation measurable in centimeters, or at most several meters, that are superimposed upon a landform, element landform or common landform. Individually these features are either too small to be delineated (and are therefore used to describe a site), or they occur grouped in a repeating pattern which are mappable (such as patterned ground features).

Morphometry

Morphometry is defined as "The measurement and mathematical analysis of the configuration of the Earth's surface and of the shape and dimensions of its landforms". It is most often applied to a geomorphic map unit to provide the quantification needed to ensure consistent application for mapping, correlation, and interpretation purposes. The following measurements or characterizations of elements of the landforms are used in the morphometry portion of this geomorphic classification system:

> Relief, elevation, aspect, slope gradient, slope position, position/landform modifier, slope shape vertical, slope shape horizontal, slope complexity, ground surface shape, landform width, microfeature relief, dissection frequency, dissection frequency class, dissection depth value, dissection depth class, drainage pattern, drainage density and stream frequency.

Not all parameters are applicable at all scales. Also, these parameters can be expressed as ranges of values, averages or means as an application of the classification.

Following are the various morphometry parameters, which can be measured or described:

Relief - Differences in elevation of a land surface or specific geomorphic feature, measured in meters or feet.

Elevation - The elevation of a land surface, measured in meters or feet from the average of the mean high and mean low tide.

Aspect - Direction in which a slope faces, measured in degrees of azimuth or cardinal direction.

Slope Gradient - The gradient of the inclined surface of any part of the Earth's surface, measured in percent or degrees.

Slope Position - A description of the two-dimensional position on the slope profile of the landform, e.g., summit or back slope.

Position/Landform Modifier - A modifier which may be used to better describe the slope position or position on a landform, e.g., upper, second, windward, distal, and landward.

Landform Width - Measurement of the width of landforms, such as terraces, floodplains, stream channels, and glacial troughs, measured in meters or feet.

Slope Shape Vertical - Refers to the vertical slope shape of the land surface, e.g., concave, linear, and broken.

Slope Shape Horizontal - Refers to the horizontal slope shape of the land surface, e.g., concave, linear, and broken.

Slope Complexity - Refers to the complexity of slope shape for a point or polygon. Simple slopes may be further described as linear/planar, concave, or convex. Complex slopes may be further described as broken, undulating, or patterned.

Ground Surface Shape - Shape of the ground surface (as opposed to the land surface), expressed as either hummocky or uniform. May be further described by identifying the microfeature influencing the surface shape.

Dissection Frequency - The number of stream channels per linear mile or kilometer, within a polygon.

Dissection Frequency Class - The frequency class for dissections in landforms with slope components.

Dissection Depth Class - The dissection depth class for landforms with slope components. Dissection Depth Value - The actual measured value of dissection depth in feet or meters. Drainage Density - The density or total length of stream channels on a landform expressed in terms of miles of channel per square mile of land or kilometers of stream per hectare of land.

Drainage Pattern - The configuration or arrangement in plan view of the natural stream courses in an area, e.g., dendritic, trellis, angulate, contorted.

Microfeature Relief - Refers to the vertical change in elevation of the land surface at scales measurable in centimeters to a few meters, or inches and feet, that are superimposed upon a larger landform and are too small to delineate on a topographic map or aerial photograph, at commonly used scales.

Stream Frequency - The measurement of the total number of stream channels of all orders, divided by the area of a basin, watershed, or polygon.

Geomorphic Generation

Geomorphic generation is a component of the classification that allows for the recognition and documents the status of more than one geomorphic type at any given location on the ground (overprinting). Overprinting occurs when two or more dissimilar geomorphic processes have operated on an area at different time periods due to the influences of climatic changes and/or tectonics. For instance, an extensive uplifted Tertiary river terrace has a mass wasting translational slide landform superimposed on it. Determination of the geomorphic generation of each landform will identify the genesis of each of the landforms, the relationship between the landforms, and the status of the process, which formed or continues to form the landforms.

A polygon or point can be attributed for "Geomorphic Generation" using the following terms:

Active - A landform which is continuing to develop under the contemporary processes that formed it, such as a floodplain, where tectonic or climatic "events", which are influencing the landform, occur on the scale of hundreds of years, and are expected to continue to occur. Examples include Holocene fault landforms, recently active mass wasting landforms, alluvial fans, and Holocene volcanic composite cones.

Dormant - A landform which evolved or developed under different geologic and climatic influences, which do not occur in contemporary times. Processes which formed these landforms could re-occur, but the cycles responsible for them occur on the order of thousands or tens of thousands of years. Examples would be glacial, periglacial and eolian landforms, which evolved in the Pleistocene, when climatic conditions were cooler and/or wetter.

Relict - Landforms or remnants of landforms which developed in previous geologic periods which remain recognizable well past the period of their development which are unlikely to renew development under contemporary conditions. Examples would include volcanic landforms, which developed in the Tertiary, which have not exhibited volcanic activity for several million years, eroded fault scarps in tectonically stable areas, and deeply incised coastal plains.

For mass wasting landforms, attributing for geomorphic generation must not be confused with either natural or human induced slope stability hazard evaluation. A mass-wasting feature, which is considered "active" from a geomorphic generation perspective, may well have a generally low natural slope stability hazard, while a "dormant" mass-wasting feature may have a high susceptibility to be re-activated by certain human activities such as road construction. Determining slope stability hazards requires complete analysis of many factors, including identification of the geomorphic process and landform, morphometry, material characteristics, groundwater conditions, bedrock structure, as well as other local factors.

Spatial and Temporal Scales

Different physical and biological processes can have dynamic interactions when they operate on the same spatial and temporal scales. In this chapter, the spatial and temporal scales are defined for estuaries by looking at the interactions between several factors that lead to variations in the stability and morphology of fine intertidal sediment shores.

Scale Interactions in Biogeomorphology

Different physical and biological processes can have dynamic interactions when they operate on the same spatial and temporal scales. Processes that act on a very small scale may appear as noise in the interactions with processes on larger scales. Their effect can be accounted for by proper averaging procedures (e.g. for turbulence). Processes that act on a large scale may be treated as slowly varying or even constant boundary conditions when studying their effects on processes on smaller scales (e.g. sea level rise due to climate change). Techniques for scale interactions are reasonably well established in geomorphology and are based on scale linkage via sediment transport. In biology however, population and community dynamics give rise to spatial and temporal structures that are not easily linked. In recent years the importance of scale has been increasingly recognized as an essential aspect of understanding the biotic and abiotic processes that affect the biogeomorphology of coastal systems.

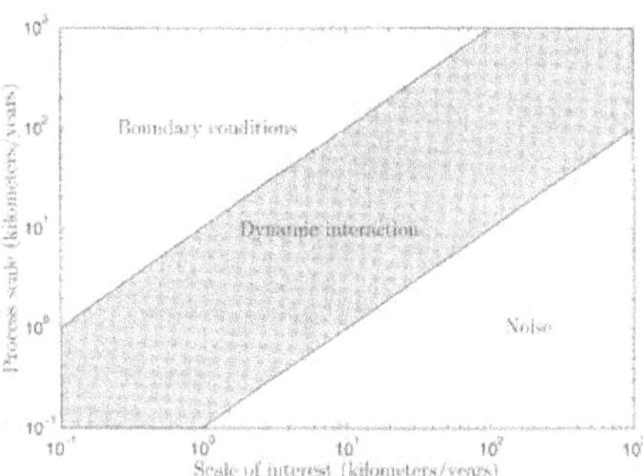

Figure: Scale concept in which influences from higher scales are described as boundary conditions, and influences from lower scales are considered noise. Dynamic interaction for processes on the same process and interest scale

Spatial and Temporal Scales Defined for Estuaries

There are marked spatial and temporal variations in the stability and morphology of the intertidal fine sediment shores in estuaries and embayments of NW Europe. This is due to the interactions between several factors:

- The spatial distribution of biota acting as ecosystem engineers in relation to intertidal height and the estuarine salinity gradient,
- The seasonal and inter-annual changes in these biota and particularly the relative abundance of bio-stabilisers and -destabilisers,
- Physical drivers determining biota abundance (e.g. temperature, salinity), and
- The frequency and intensity of erosive forces (e.g. wind-wave activity, storms, rainfall and fluvial flow).

For example, field studies in the Humber (England) and Westerschelde (Netherlands) have shown significant temporal-spatial changes in intertidal sediment erodability and morphology which reflects the interaction between physical processes of erosion (tidal currents and waves) and the abundance of biological stabilisers (benthic diatoms) and destabilisers (the clam Macoma balthica). Following cold winters there is an increase in the density of the bioturbating clam, which causes an increase in intertidal sediment erodability. Current and wave induced erosion of the mudflat, elevates suspended sediment concentrations, which are then transported up the shore on the flood tide and onto the salt marsh where there is enhanced sediment accretion. Conversely, when clam densities are low and grazing pressure low (usually following a series of mild winters), there is enhanced stabilisation of the mudflat by benthic diatoms which leads to a reduction in sediment resuspension into the water column and an order of magnitude lower rate of accretion on the salt marsh. This provides an interesting parallel between man's stabilizing influence on long-shore sediment transport (i.e. use of groynes), and biota's influence on up-shore sediment transport. Namely, stabilising sediment in one area is likely to reduce transport and accretion in another area.

Coupling of Mudflat to Saltmarsh

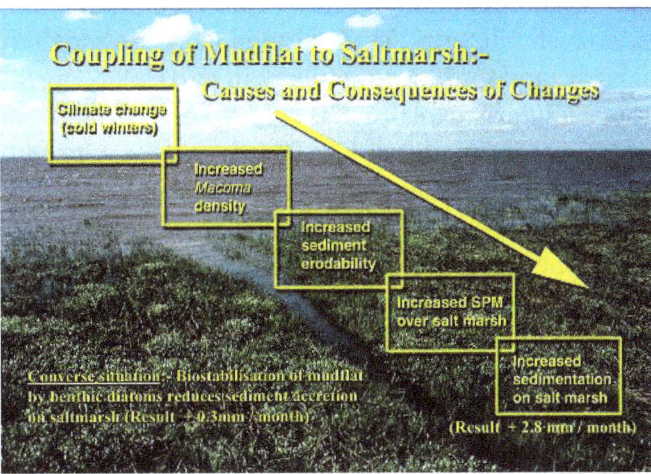

Figure: Coupling of scales for a mudflat and a saltmarsh system. From climate change to increased sedimentation on salt marshes.

Research suggests that predicted climate change (i.e. global warming with associated sea level rise) is likely to induce a shift towards increased bio-stabilisation of intertidal sediments (micro-organisms and macroalgae) and as a consequence reduce sediment accretion on salt marshes. However, climate models also predict increases in the frequency and intensity of storms, but at present it is unknown whether this will result in net erosion or net accretion on upper shore salt marshes. Both situations have been recorded and the net result may be dependent on various factors including the sediment supply and grain size, the shore profile, and the orientation of salt marsh relative to prevailing winds.

Biogeomorphology

Biogeomorphology (also commonly referred to as ecogeomorphology) is the branch of geomorphology, which focuses on the interactions between ecological and geomorphic processes. As such, it covers a very wide range of subject matter, from the mutual interdependence of microorganisms and weathering on bare rock surfaces to the interactions between forest cover and fluvial dynamics within whole catchments. Biogeomorphic interactions occur in all terrestrial environments, from the hyper-arid zone to the wet tropics, and are also heavily affected by human impacts on the environment. Thus, biogeomorphology has a very broad canvas, one, which has global, reaches and covers many different scales.

Themes for a Complex Biogeomorphology

Multiple Causality

Multiple causality is the first of four themes underlying complex systems in biogeomorphology. Multiple causality has long been acknowledged in ecology and geomorphology through the process–form interaction. Ecologists and geomorphologists have historically, however, relied on one or the other of these two causal positions rather than circularity to explain geographic observations. To an extent, this preference for a unidirectional interpretation is a compromise made in the interest of methodological tractability and clarity in language rather than any conceptual shortsightedness. Yet for ecologists, and for biogeographers in particular, the distribution of vegetation across a landscape is often singularly explained as an outcome of disturbance patch dynamics. In this case, process shapes vegetation patterns. Fire, floods, and gravity-induced slope failures create patches of vegetation of different age and composition. The distribution of vegetation is also a response to the underlying structure of environmental gradients of soil, water, and nutrients. In this case, form is invoked as the causal agent. Like the division in geomorphology between form-based, historical narratives versus process-oriented explanations of geomorphic change, biogeographers are split between a process or form explanatory framework. Yet for some Earth surface systems, biogeographic patterns may better reflect the interaction of disturbance processes and recovery along environmental gradients.

For a longer period of history, and perhaps more openly, geomorphologists have understood that the relationship between landform and process is axiomatic. Geomorphic processes shape landforms, and these landforms shape geomorphic processes. Geomorphologists have also periodically over-relied on one of these two explanatory frameworks, process or form, each with its own underlying assumptions and favored methods. Process-based studies are reductionist and seek to apply mechanistic rules to explain landform patterns. In this perspective, deterministic interactions at the local scale are the causal factors of relevance to understanding Earth surface systems. Form-based interpretations of landscapes invoke larger time scales and more conceptual, less experimental modes of explanation. In this view, geomorphic landscapes can be explained by contingent geographic factors and particularistic historical narratives. The reductionist models of plant succession, proposed by Connell and Slatyer, are ecological counterparts of the deterministic, process-oriented vein in geomorphological inquiry. Likewise, the ideas about succession initiated by Clements, still resound as an example of narrative, form-based thought in ecology.

A complex systems biogeomorphology encompasses the permutations of this process–form causal interplay. In this view, the process–form interaction becomes the central epistemological (and methodological) strand linking geomorphology and ecology. As an illustration, consider the disturbance by fire and mass wasting in steep forested terrain. Topographic variation can influence ecological processes, such as disturbance by fire, through slope aspect or shape. Conversely, disturbance by fire can exert an influence on surface erosion and slope stability via its impacts on vegetation. Ecological structure, specifically the type and patterning of vegetation, influences erosional processes and can potentially modify downslope sediment transport. When of sufficient magnitude, slope failure can modify the spread and extent of disturbance by fire by creating fuel breaks and a patchy fuel mosaic.

When this full circularity of biogeomorphology is acknowledged, any ranking of the causal precedence of landform characteristics, the associated geomorphic processes, or vegetation, becomes tenuous.

From the perspective of complexity theory, the distributional patterns we observe are an irreducible property of the dynamics of a system. Efforts to untangle the importance of one causal agent over another become scale dependent. What becomes more relevant is how this interaction of process and form plays out across time and space. Specifically, to what extent does this multiple causality become self-organizing, or recursive? When we say a system is recursive, the emergence of system properties can act as a constraint upon future development. Levin lists three conditions for the development of the recursive interactions that define a self-organizing complex system:

1. A sustained diversity and individuality of components;
2. Localized interactions among these components;
3. A selecting process that fosters a subset of these components for replication and enhancement.

In a study of how flood disturbance shapes the richness of patterns in riparian plant species and diversity within two watersheds in central California, Bendix echoed the concept of recursivity when he speculated 'how much is the pattern of diversity affected by the variation of species composition through a watershed, with its concomitant impact on disturbance vulnerability.' Recursivity in this passage denotes how disturbance by floods and vegetation can potentially shape and be shaped by each other. This interaction can exert a sorting effect, a self-organizing regulation, such that disturbance exposure and vegetation responses cannot be decoupled from the patterns of diversity. Other biogeomorphic systems also exhibit recursivity. The variability of tree compositions on non-mountainous glacial landforms can be explained by substrate properties. These vegetation patterns reinforce soil contrasts among landforms by producing litter of different nutrient composition. This in turn reinforces the vegetation contrasts and the differentiation of substrate properties.

Such feedback scenarios are common in ecological and geomorphic systems, but the characteristics of the geographic signatures are less examined. To comprehend how recursivity has relevance for our understanding of biogeographic patterns, one must consider its effects upon environmental heterogeneity. Environmental heterogeneity is strongly associated with the distribution of species and the patterns of biodiversity. Heterogeneity can also emerge, however, from the recursive interactions of organisms and the geomorphic setting. Although the positive correlation between geomorphic heterogeneity and species diversity has been recognized, the degree to which heterogeneity is a cause and an effect of its relationship with biota has not been well investigated. Usually only an external source of the heterogeneity, such as disturbance or physiography, is invoked. The multiple causality in fast biogeomorphic systems, such as riparian zones and dunes, may be productive settings to partition these sources of heterogeneity and delineate how they are intertwined with the patterns of species diversity.

Ecosystem Engineers

Some species exert a disproportionately stronger influence on overall ecosystem structure and function than others. Most ecologists probably regard trophic links as the main mechanism by which these keystone species exert influence. From a geomorphological point of view, however, the material impacts of biota may be just as important. Through the construction of habitat or modulation of the movement of sediment and water, individual organisms and species can have large effects on entire communities. The beaver is probably the best-known example. Geomorphologists have long acknowledged the geomorphic agency of beavers as well as other organisms, from elephants and plants to invertebrates and cyanobacteria. But how these species, known in the ecology literature as ecosystem engineers, are incorporated into biogeomorphic frameworks of complexity has not been broadly acknowledged.

Evidence from a wide variety of ecosystems indicates that many organisms may stabilize habitat against erosion by wind or water. Plant roots, biotic crusts, and biofilms can increase soil and sediment surface stability. Riparian vegetation affects the morphology of channel cross-sections and floodplains through its ability to bind sediments. Mosses stabilize potentially mobile sediments in coastal dune systems. Aquatic invertebrates can modify the shear stresses of sediments.

In addition, ecosystem engineers can facilitate species coexistence by providing habitat for themselves and other species. Dune landforms, formed from positive feedbacks between vegetation growth and sediment entrapment, create protective microhabitats in the lee that increase the richness of local species. Spartina alterniflora, a common grass of salt marshes and beaches, facilitates the establishment and persistence of cobble beach plant communities by stabilizing the substrate and enabling seedlings of other species to emerge and survive. By increasing the topographic complexity of benthic habitats, aquatic macroinvertebrates alter patterns of near-bed flow such that the feeding success of individuals is enhanced and species coexistence is augmented. Some invasive riparian species, such as Ligustrum sinense (Chinese privet) or Tamarisk spp. (salt cedar), may facilitate persistence by altering the patterns of sediment erosion and deposition along riparian landforms.

By stabilizing substrates, enhancing weathering processes, providing habitat, and promoting facilitative relationships, ecosystem engineers introduce some of the requisite nonlinearity for dynamic activity in complex systems. Effects initiate time lags, legacies, and slowly appearing indirect effects that can ripple throughout a landscape. Even the nonliving products of organisms, beaver dams, coralline skeletons and woody debris, can influence a system long after the organism has died. These biotic imprints may communicate across temporal and spatial scales to entrench and shape community dynamics. Nonlinearity induced by ecosystem engineers can potentially lead to historical path dependency, whereby future interactions are constrained by initial effects. A corollary of path dependency is the potential for multiple stable states.

Organism-mediated nonlinearity can also arise through the modification of geomorphic thresholds. During the intervals between external disturbances, ecosystem engineers can modify geomorphic thresholds and subsequent responses to extrinsic events. For example, with the plant-mediated formation of dune landforms, inland habitats in the short run are less likely to 210 J.A. Stallins / Geomorphology 77 207–216 be exposed to disturbance from overwash generated during passing storms. In wetlands, vegetation may alter sedimentation and fine-scale elevation, thereby modifying the patterns of inundation and disturbance exposure. Following ideas from landscape sensitivity, biogeomorphic interactions modify the temporal and spatial distributions of resisting and disturbing forces. These examples suggest that ecosystem engineers have the propensity to introduce a nonlinear relationship between disturbance forcing event frequency and subsequent exposure. In other words, because of biotic modifications of geomorphic thresholds, a simple tit-for-tat relationship between the event forcing disturbance and exposure cannot always be expected. This interaction is similar to Schumm's idea of complex responses in geomorphic systems, whereby different responses can develop from the same conditions of perturbation. In ecology, resonance and attenuation theory similarly posit that intrinsic processes may interact with periodic extrinsic fluctuations so as to reinforce or dampen effects. The extent of these nonlinear influences is dependent upon the durability of constructs, artifacts, and impacts in the absence of the original engineer.

Ecological Topology

Parker and Bendix stressed the need for a better understanding of how biogeomorphic interactions vary geographically. A major constraint to this task is how to visualize the multiple scales and cause–effect interactions that define complex biogeomorphic systems. Ecologists working in the area of complexity theory have developed conceptual and quantitative models for visualizing this complexity in space and in time. These models work from the premise that natural and human-coupled systems may exhibit a range of assembly states depending upon the initial conditions and the extent of the recursive, nonlinear interactions. The shifts between these assembly states can exhibit threshold-like responses and be triggered by humans or by non-anthropogenic environmental or ecological change.

The self-reinforcing, assembly states that can emerge from these nonlinear interactions go by a variety of synonyms: attractors, stability domains, domains of scale, or process domains. These entities demarcate the boundaries in time and/or space over which process and form reinforce another. Within each stability domain, a small set of species and abiotic processes mediate structure and function, and exert some control over reproduction. The regions bridging stability domains are described as transient states or phase transitions. These locations exhibit weaker recursive properties, and a higher turnover in the arrangement of feedbacks among components. This geometry of stability domains and phase transitions is often abstracted as a fitness landscape of hills and valley, giving rise to a 'lumpy' or 'granular' ecological fabric, or topology. One might consider the parameters of this topology an extension of phase space, a concept originating more directly from complexity theory.

This ecological topology has been used to conceptualize the spatiotemporal domains of causality in biogeomorphic systems. Feedbacks between regular, predictable forcings of high flows of water and organisms in riparian zones may become a mechanism for defining distinctive domains of recursive, biogeomorphic cause and effect. Shallow lakes can flip through time between a nutrient rich and a nutrient-poor stability domain in response to changes in sediment and nutrient inputs. Barrier island stability domains can be defined by biogeomorphic propensity to reinforce or damp overwash exposure. The concept of stability domain has also been invoked for rivers, coral reefs, and rangelands. In each of these examples, ecosystem engineers directly or indirectly mediate flows of matter, energy, and disturbance, setting up recursive, nonlinear feedbacks that confer a degree of persistence to a particular organizational state or domain.

A lumpy ecological topology can emerge when a few key recursive structuring processes establish a temporal frequency that entrains, or captures, other processes and forms. This process is similar to slaving, a concept with geomorphological leanings. In slaving, variables with disparate time scales, when nonlinearly coupled, can develop an asymmetric relationship. Fast variables become entrained or slaved to slow variables and lose status as independent dynamical variables. For example, along sandy coastlines, the fast motion of sand grains can become slaved to the slower motion of sand dunes. In turn, sand dunes can be slaved to the migrations of the shoreline. Dune plants are slaved to these motions, but they also exert a degree of control on the process. The extent of this control could be expected to vary as a consequence of dispersal processes, the magnitude or intensity of sediment transport, as well as the degree to which the temporal scaling of vegetation dynamics and geomorphic processes overlap. Hierarchy theory in ecology similarly recognizes that the coupling among processes operating at different scales can vary from strong to weak.

Stability domains may be more apparent in some biogeomorphic environments than others. Where geomorphic processes operate very slowly, biogeomorphic interactions driven by vegetation may be less visible. Steep gravitational gradients may complicate domain structure. Because of linear geometry, rivers may require a combination of continuum and lumpy approaches. Where disturbances are large and return times frequent, domain structure may not develop. Topology for many biogeomorphic systems may ultimately be constrained by sediment budgets. Lastly, the distribution of domains across a landscape should reflect deterministic and contingent conditions. In other words, topology should be a function of the contingent underlying variation in physical variables, as well as the propensity for species and the related environments to self organize.

Ecological Memory

Ecological memory is another prominent theme that ties geomorphology and ecology together under the framework of complexity. Ecological memory encompasses how a subset of abiotic and biotic components are selected and reproduced by recursive constraints on each other. This recursiveness has the potential to become canalized through time, whereby it is encoded in organisms, and to an extent the immediate environment.

Disturbance can become encoded and reinforced in the abundance and spatial pattern of vegetation and topography across a landscape. In the longleaf pines forests of the southeastern U.S., fire can become a replicable process encoded in the structure of a biological community as a result of past environmental conditions, the distribution of topographic and soil variables, and subsequent selection on populations. In mountainous-forested terrain, slope failures form firebreaks that interrupt fuel connectivity and limit the size of fires. In turn, the vegetation and landform characteristics that emerge following these disturbances (such as the abundance of fire-enhancing or slope-stabilizing plant species, the density and composition of seed bank, slope angle and stability, soil permeability and nutrient content) may or may not reinforce the replication of these landscape processes. Memory can also occur on smaller, more discrete scales. The location of trees may be influenced by the past location of trees, via the effects upon soil properties.

This concept of ecological memory is strikingly visible in vegetated coastal dunes, systems in which history is often perceived to be constantly erased. In barrier island dune systems, the frequency and spatial extent of overwash exposure initiated by passing extra tropical storms are dynamically encoded in the interactions among this disturbance agent, species abundances and topography. Species abundances reflect the local disturbance regime not through passive adaptation, but by constructing and reinforcing topographic niches in light of the historic frequency at which disturbance forcings have occurred. In other words, by modifying topography in the periods between storm-forced overwash disturbance, dune species interdependently facilitate the historically prevailing patterns of mobility of the surface sediments and species abundances in a positive feedback. At locations where storm forcings of overwash disturbance are more historically frequent, plant species well adapted to sediment burial but lacking growth forms that enhance dune-building may promote abundance and persistence by contributing to a low-profile topography that lowers the resistance to future overwash exposure. Where disturbance is infrequent, dune-building plant species reinforce presence in the landscape by contributing to a high topographic roughness that damps overwash exposure. At both locations, the historic interaction of dune plant species and the local patterns of sediment mobility 'remembers' or perpetuates the topographic habitats and disturbance processes for these species.

Ecological memory does not denote the organismal idea of a preordained final successional structure. Biotic and geomorphic components of a landscape can exert some control over reproduction, but memory in this sense is contingent and open-ended. Sequential species replacement can occur, but superimposed upon this may be the effects of ecosystem engineers. Their subsequent responses to and effect upon environmental variability, whether novel or historically prevalent, have the propensity to influence how biogeomorphic interactions play out and create geographic patterns.

Coastal Biogeomorphology

Coastal zones are where some of the most obvious types of biogeomorphological landforms and processes are found. Biogenic habitats such as eye-catching coral, serpulid and worm reefs; mangrove forests and saltmarshes are constructed and crusts of rocky shore species like barnacles, oysters and mussels grow at the coast. These species also serve a bio protective function, where they buffer waves or trap sediment. Many species are busy grazing for food or drilling into rocks to make their homes, bio eroding the rocks and shaping the shore at the same time.

Biogeomorphologists have an interest in understanding how biogeomorphic species interact with their landscape to create, alter and sustain landforms from a range of external forces such as waves, thermal expansion and contraction, wetting and drying, and the action of salts. These organisms can increase (i.e. bioeroders) or decrease (e.g. bio protectors) the rates of other important weathering processes – it is these interactions that shape, modify and sustain coastal landforms. Research often involves a combination of field trials, laboratory experiments and microscopy.

Coastal biogeomorphology researchers often help solve real-world problems. They do this by:

1. Identifying the bioprotective value of species. Saltmarshes provide a buffer against waves, which means that seawalls behind them don't have to be as big, which costs less. Coastal managers now take this into account when designing new coastal defence structures.
2. There is a growing requirement for coastal managers to work with natural processes when managing soft coasts at risk of erosion. Biogeomorphologists often help do this as part of saltmarsh or mangrove restoration programmes.
3. New research is demonstrating that biogeomorphological science can be used to help ecologically enhance hard coastal structures. This is showing that you can (a) engineer structures to increase the rate of colonization and (b) that some species protect the coastal infrastructure from deterioration.

Climatic Geomorphology

Climate geomorphology is a branch of geomorphology which deals with the effects of climate on geomorphological processes and consequently on the character of land-forms. It has included the identification of climatically controlled zones and has attempted to define provinces with distinctive denudational processes. Geomorphic mechanisms vary in type and rate according to the particular climatic zone in which they function. Landforms produced from these mechanisms will be different from region to region and will reflect the dominant climate.

The New Classification

Morphoclimatic zones were determined by the natural divisions resulting from atmospheric circulation patterns. The system of seven belts of differing morphological processes corresponds roughly to Flohn's system based on the climatic influence of four major pressure belts with "steady" climates and three adjacent belts with "alternating" climates.

The postulation of a causal link between climatic classifications and climatic geomorphology, along with an associated definition of zones according to specific threshold values, which are easily obtained and globally applicable, give us a climatic classification system useful for geomorphological purposes. Although it is an "effective" classification, it also contains significant "genetic" elements. A further innovation is the geomorphological orientation, using the factors of climate, principal morphodynamic processes, and typical recent land forms to denominate the zones in a homogeneous form.

Some areas are exceptions. For example, the polar ice caps inhibit direct atmospheric action in land surface formation. However, this factor is dealt with via creation of a subglacial zone. The zone limit in the direction of the equator is the extent of permanent ice cover and is defined by the 0 °C-isotherm for the warmest month.

Mountain regions, which are azonal in their areal distribution, should also be excluded; the hypsometrical change in landforms is very important in this case. Since adibiatic deserts owe their formation exclusively to surrounding mountains and are not situated at a distinct latitude, they should be placed in a separate category.

Allowing for these exceptions, we are left with six zones of differing recent morphodynamic activity. The areal distribution is given for each, followed by an analysis of typical landforms and processes. Finally, the limits of the zones are defined according to meteorological threshold values.

1. The Subpolar Zone of Solifluction and Frost-Shattered Debris

Areal Distribution

This zone includes northern Canada; Alaska; coastal regions of Greenland; the extreme northeastern parts of Europe; and the whole of Siberia.

Characteristic Landforms and Processes

One of the most striking features of this zone is the cover of debris ranging in size from small fragments to large blocks. Genesis is attributed to an efficient mechanical weathering caused by the expansion of water as it freezes in rock cracks.

In general, the smaller fragments are characteristically sorted to form various kinds of frost-patterned ground. The freeze-thaw cycle, with its associated changes in volume, causes the coarser material to migrate slowly to the surface.

Because frost-patterned ground is formed by periodic, morphologically efficient ground frost, this type of landform is more highly developed in regions of permafrost.

The solifluction process is very common and enhanced by ground frost. During the summer, surface ice melts and infiltration of the water is prevented by the permafrost layer. Soils above the permafrost layer become saturated with water.

Valleys have broad, debris-covered bottoms. During the cold season, when water is scarce, the bedrock is shattered by frost and loosened; creating favorable conditions for erosion due to the sudden high volume of water during snow melt ("ice-rind effect").

Many of the phenomena occurring in adjacent regions with tundra vegetation are similar to those in regions without vegetation cover. Nevertheless, frost action remains the principal factor and some landforms, such as pingos, are due to the existence of permafrost.

In considering typical zonal landforms and processes, it is not advisable to introduce a separate subpolar zone with tundra vegetation, which would be of equivalent importance to the other six zones. However, its classification as a sub-zone within the subpolar region, characterized by weaker morphodynamic processes, is reasonable and a practical solution.

A final note should be made on the landforms in areas of retreating permafrost, particularly those caused by solitabetion 1. They should be used as criteria of zone definition only if they are caused by recently occurring cyclic processes, which are not due to a major change in climate.

Meteorological Threshold Values

The above description of landforms and processes has clearly shown that the freezing point is of particular significance; the freezing and thawing of water has many important consequences for landform genesis. Various effects are of particular note. For one thing, the oscillation of temperatures around the freezing point is of primary significance. Frequency of freezing and thawing, together with time, are the decisive factors in the widespread shattering so typical of periglacial regions; the formation of patterned ground is also substantially affected by this frequency factor.

The frequency of freeze-thaw cycles is dependent on microclimatic conditions, since frost must occur frequently enough and penetrate deeply enough to be morphologically effective. Therefore the appropriate meteorological threshold value will be an approximation based on some relationship such as that between the number of morphodynamically effective frost cycles and standard air temperature measurements. The maximum number of effective cycles occurs when the annual mean temperature is between 0 °C and -2 °C.

As mentioned previously, the presence of permafrost is considered the second most important element. Since indices become unwieldy and inaccurate when all significant factors have been integrated, an exact definition of permafrost using a meteorological threshold value is not possible. For these reasons, the definition of limits based on an annual isotherm is preferable and corresponds closely to the outer limit of discontinuous permafrost in the direction of the equator. The annual isotherm of -1 °C is generally accepted for this purpose.

The annual isotherm of -1 °C gives a fairly accurate indication of a high frequency of frost cycles and discontinuous permafrost.

2. The Moderate Zone of Modicofluvial Action and Relict Forms

Areal Distribution

This zone includes southern Canada; certain areas of the U.S.A. east of the Rocky Mountains; Europe, with the exception of the Mediterranean region; the northeastern part of China; the southern tip of South America; the southeastern tip of Africa; southeastern Australia; and parts of Tasmania and New Zealand.

Characteristic Landforms and Processes

Since we are dealing here with present day landforms and processes, the method is not appropriate for this zone. Because recent morphological activity has been very weak, inherited landforms have not been fundamentally altered. Although fluvial erosion has produced some alteration of large valley bottoms, this process is also moderate in its effect.

Soil erosion is one of the most intensive processes in this zone; however, this erosion is largely anthropic in origin and agriculture is largely responsible for partial destruction of the protective vegetation cover.

Meteorological Threshold Values

Since meteorological factors of land formation are of moderate effect in this zone, they do not contribute to a reliable identification.

For this reason it is appropriate to include in this zone all those areas not characterized by the active morphodynamic systems of the adjacent zones. The pole ward limit is therefore determined by means of the annual isotherm of -1 °C, and the equatorward limit by the corresponding threshold value of the subtropical zone, which remains to be discussed.

3. The Subtropical Zone of Slope Wash and Seasonal Rivers

Areal Distribution

This zone includes northern parts of west coast U.S.A.; an area east of the Rocky Mountains at the same latitude; Mediterranean Europe; some coastal regions of North Africa; western Asia; China north of the Changjiang; the area surrounding the La Plata region; the southwestern tip of Africa; southern parts of Australia; parts of Tasmania.

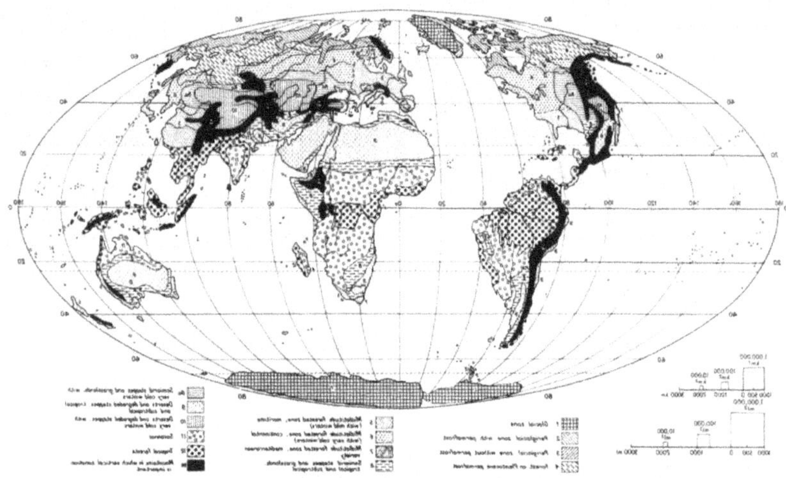

Figure: Morphoclimatic zones of the earth

Characteristic Landforms And Processes

The characteristic gullying of slopes in this zone is due to the intensive slope wash created by the seasonal distribution of precipitation.

Since there has been anthropic intervention in ecological processes over a long period of time, it is difficult to evaluate whether the morphodynamic processes forming large valley bottoms with layers of debris are principally natural or pseudo-natural in origin. The fact that regions with large monthly differences in precipitation are generally subject to a high amount of fluvial erosion supports the hypothesis that the formation of the typical valley bottoms is caused by a characteristic seasonal variation in rainfall.

Frequent landslides, however, must be attributed for the most part to anthropic intervention. These particular forms of mass wasting do not therefore enter into a discussion of factors of climatic geomorphology, particularly since they are more often associated with specific pétrographie or stratigraphie conditions.

Meteorological Threshold Values

Defining the limits of the subtropical zone in terms of the usual meteorological mean values is not very appropriate, since the critical factor is the seasonal variation in precipitation.

The question is which of the two elements of pronounced drought or heavy rainfall is more significant. Although pronounced drought inhibits the growth of vegetation and delays soil formation processes, it has no direct geomorphological effect. On the other hand, gullying by slope wash and periodic runoff points up the importance of periodic precipitation; it facilitates the transport of rubble on slopes and the deposition of debris.

The wettest month, and not the driest, should therefore form the basis for the index; the following is then appropriate.

The seasonal basis of active development processes and landforms in the subtropical zone of slope wash and seasonal rivers can be expressed by means of a simple index.

In this zone belong all regions where, for a summer rainy season, precipitation during the wettest month is double the monthly average, or, for a winter rainy season, precipitation during the wettest month is more than one-and-a-half-times greater than the monthly average. The index is therefore:

$$\text{Summer Rainy Season} \quad \frac{\text{Precipitation} - \text{wettest month}}{2 \text{ X Average monthly precipitation}} = 1$$

$$\text{Winter Rainy Season} \quad \frac{\text{Precipitation} - \text{wettest month}}{1.5 \text{ Average monthly precipitation}} = 1$$

For the subtropical zone of slope wash and seasonal rivers, the resulting figure is greater than 1.

The need to establish causality has been met through a seasonally-based expression. The factors of 2 and 1.5 could not have been reached by theoretical means alone, but rather by comparing the areal distributions of characteristics in the zone to be defined with corresponding characteristics in other geomorphological zonations.

In addition to periodicity of precipitation, there must also be present the minimum amount of rainfall necessary to sustain the characteristic processes. Thus the precipitation threshold of desert zones has priority in defining the limit between the subtropical zone and desert zones. Moreover, a thermal threshold value is also necessary when the subtropical zone borders on the tropical zone, since the latter is also characterized by periodic rainfall. This threshold value should be welldefined in dealing with the tropical zone, because the capacity of its morphodynamic system to shape land surfaces is significantly greater.

A satisfactory definition towards the moderate zone requires one other threshold value, since temperatures around -1 °C set off the morphological mechanisms of the subpolar zone. Since it is very effective in land formation, this mechanism can modify landforms even if only of short duration in the annual cycle.

A value based on a period of low temperatures is appropriate, since low temperatures can produce a change over an entire winter period, which is three months according to meteorological definition. Therefore, if three months of low temperatures serve as the criterion for frost-action efficiency, a three-month period of temperatures below -1 °C is used to define the poleward limit of the subtropical zone, in conjunction with a seasonality index.

The present-day morpho-climatic zones. 1. Glacier zones. 2. The sub-polar zone of excessive valley-cutting. 3. The taiga zone of valley-cutting (within the permafrost realm). 4. The extra-tropical zone of retarded valley-cutting. 5. The subtropical zone of mixed relief development — mediterranean realm. 7. The winter-cold arid zone of surface overprinting (transformation), mainly by glacis and pediments. 8. The warm arid zone of plains preservation and traditionally continued planation, mainly by fluvio-eolian sandplains. 9. The peritropical zone of excessive planation. 10. The inner tropical zone of partial planation.

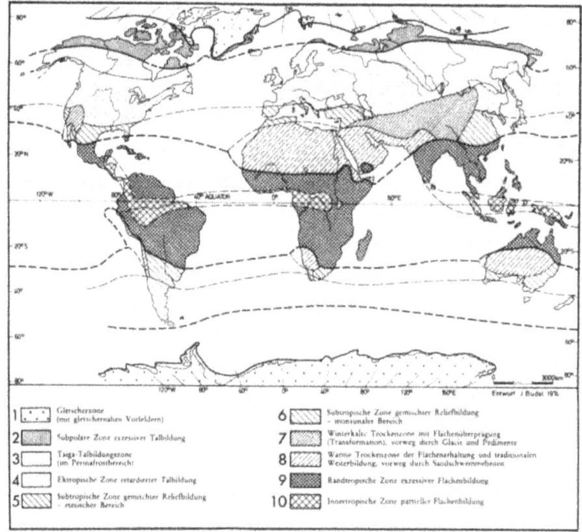

Figure: The present-day morpho-climatic zones.

4. The Paratropical Zone of Eolian Action and Dry Debris

Areal Distribution

This zone includes the desert regions of Southern California; the Sahara, Somalia, Arabian, and Great Indian Deserts; the Atacama Desert; the Namib and large parts of the Kalahari; and the desert regions of central Australia.

Characteristic Landforms And Processes

Vegetation cover, which would reduce the effects of air temperature on soil and rock surfaces, is sparse or non-existent in semi-desert and desert regions ; differences between day and night temperatures are often considerable. These regions therefore constitute a second zone where physical weathering is highly efficient. Over time there have been generated great amounts of sharp-edged debris similar to that produced in the subpolar zone by frost-shattering.

The products of this characteristic weathering are also present at the base of steep slopes, where debriscovered areas mark the transition between mountains and closed depressions filled with fine-grained sediments. The origin of these pediments is under dispute. Budel's explanation is the simplest and most plausible: tropical desert relief is a remnant of an "excessive planation of the rock surface in a long, humid, pre-Pleistocene age". Pediments are therefore mostly relicts and irrelevant to our present discussion. On the other hand, dry debris, which is recently formed, should be taken into consideration, since it has been produced by the arid morphodynamic system and it overlays the fossil landforms.

The morphodynamic efficiency of periodically active water is considerable. Rare but violent rainstorms cut groove systems and generate strong erosional activity in the wadis. Traces of closed-depression drainage can be found as progressively finer sediments in the dry lakes of alkali flats. The difference from the formation of valley landscapes in other regions is that this lesser amount of rainfall is not of sufficient magnitude to cut a continuous fluvial system or to fill the hollows with material.

In desert regions, special attention should be paid to eolian processes and the formations they produce. Landforms produced by wind action can be divided into two basic groups: those created by deflation or corrasion and those caused by accumulation. Deflation and corrasion work in tandem; serir, hammada and reg are major landforms caused by removal of finer particles and exposure of rock surfaces; blowouts should also be mentioned.

The major results of eolian accumulation are the different forms of dunes. They all require the same general conditions for formation : sufficient activity of the wind and the presence of sufficient amounts of sand, which has been produced by mechanical weathering and partly shifted by fluvial action.

Meteorological Threshold Values

Given this inventory of processes and landforms, two factors are to be considered : first, wind action ; second, a pronounced dryness. The dryness gives rise to specific weathering processes and characteristic fluvial action.

The wind factor is not a zonal phenomenon ; in fact, tropical desert zones are relatively calm. However, the efficiency of eolian action is dependent on precipitation ; low precipitation inhibits vegetation cover and therefore protection from wind erosion. What is to be determined is the degree of aridity necessary to produce characteristic processes and landforms of the paratropical zone.

By itself, rainfall data are inadequate. The fact that evaporation increases with increasing temperatures cannot be neglected, since this is a controlling factor in the growth of vegetation. Because vegetation, in turn, has a significant effect on eolian activity, this relationship is very important. Therefore temperature should also be incorporated and we should refer to indices of aridity.

An aridity index serves to show when the amount of evaporation equals that of precipitation. It is appropriate for use in causal determination towards the subtropical zone, which is characterized by periodic runoff. However, in the paratropical zone, any runoff will quickly cease because the region is "arid", i.e., the possible rate of evaporation is higher than that of precipitation. The question then lies in choosing the best index for our purposes.

Many aridity indices have been examined for their utility to geomorphological studies; some have been designed for biogeographical purposes only and others lack a causal link with morphodynamic processes because they incorporate atmospheric moisture, saturation of vapor pressure, etc. Moreover, these complex indices with extraordinary meteorological parameters are not useful for global calculations. Therefore an index combining only temperature and precipitation data is applied.

For these reasons the discussion is limited to whether an index of annual values or one based on the number of humid months following de MARTONNE's ideas, is more appropriate for geomorphological purposes. Lauer himself has called his method a "climatological foundation in numbers for the vegetational landscapes of the tropics"; his index is therefore principally of biogeographical significance. Kôppen's index relies more closely on the amount of annual precipitation available for morphodynamic processes; climatic stations with the same number of humid months can receive varying amounts of annual precipitation. However, as regards morphology, absolute quantity and intensity are the critical factors. When rainfall is violent and episodic, the effect of evaporation on the reduction of runoff becomes less crucial and therefore of less importance to the morphodynamic system.

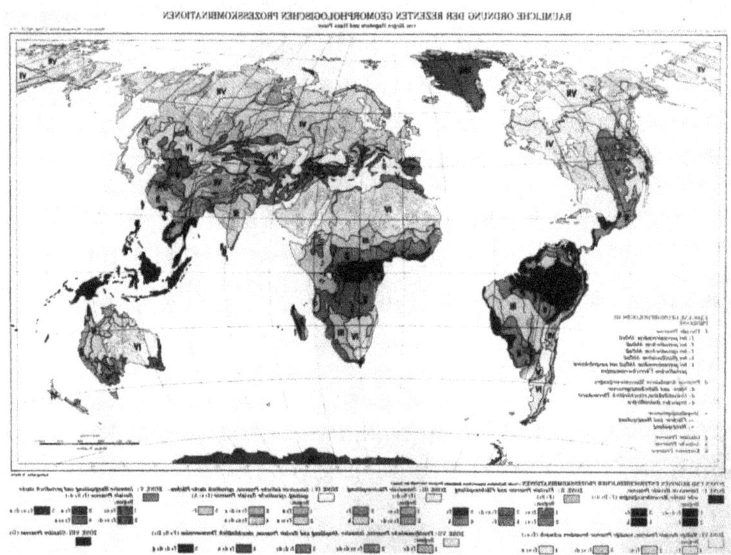

Figure: Spatial distribution of the combinations of recent geomorphological processes

Spatial distribution of the combinations of recent geomorphological processes. I. Most intense fluvial processes, very strong mass movements (Fi, Di, ds. II. Fluvial processes and sheet wash (Fi, Si). III. Most intense sheet wash (U. Si, di). IV. Most intense eolian processes, episodical strong sheet wash and episodic fluvial processes (f3, Si, A). V. Intense slope wash and periodic strong fluvial processes (f2, S2, di). VI. Moderate fluvial processes, other processes especially weak (fi, S2). VII. Cryodynamic processes, including thermoerosion, intense slope wash and fluvial processes (F 2, S2, D 2). VIII. Glacial processes (G). f. Fluvial processes: fi , by perennial runoff; f2, by periodic runoff; (3, by episodic runoff; (4, by fluvio-glacial runoff; fs, by perennial runoff with periodic inundations, s. Wash processes: 81, sheet wash; 8a, slope wash. d. Mass movement: di, falls and slides; d2 gélifluction; d3, tropical solifluction. k. Solution (karst). g. Glacial processes, a. Eolian processes.

By combining annual precipitation with temperature, Kôppen's index takes an intermediate position between a simple precipitation measurement and the use of isohygromenes. The former is related to periodic fluvial processes, the latter to eolian processes controlling vegetation.

Therefore Kôppen's aridity measure is used to define the poleward limit of the paratropical zone.

Despite its simplicity, but also because of it, this definition is satisfactory, since even more sophisticated calculation by computer with integration of sixteen variables has produced similar results. On the other hand, when dealing with regions close to the pole, SCHREIBER's parabolic aridity equation would have to be applied.

5. The Tropical Zone of Sheet Wash and Inselbergs

Areal Distribution

This zone covers the extreme southern parts of the U.S.A.; major portions of the Caribbean region; parts of Peru, Ecuador, Venezuela; Bolivia; Paraguay; northern Argentina; large portions of Brazil excluding the Amazon region ; the area of Africa between the dry regions of the Sahara and the Namib/Kalahari, except the equatorial rainforest regions; India; parts of Sri Lanka; southeast Asia and China south of the Changjiang; northern and eastern parts of Australia; parts of the islands in southeast Asia.

Characteristic Landforms and Processes

The first fully developed theory of morphogenesis for this zone was produced by BUDEL. Strong chemical weathering of bedrock and limited vegetation density due to periodic humidity are regarded as the principal contributing factors. While a high intensity of chemical weathering produces a soil surface, which can easily be removed, bedrock is continually decomposed by the combined action of high temperatures and periodic humidity; thus the rock surface remains to some extent covered by weathering products ("double planation surface").

Erosion occurs during the rainy season, after dried soils have become saturated and further precipitation runs off in rills which pick up the smaller particles and carry them short distances, often only as far as the next barrier. Since the flowing water follows no fixed path and can, during subsequent showers, pick up soil particles at other points across the slope, the total effect of erosion is spatiai and not linear. The effect is further reinforced by fairly frequent sheet flooding during downpours. These processes account for the widespread peneplains in this zone.

The way drainage occurs is notable. Because of the thick covering of chemically weathered soil, there are no abrasive tools available and the possibility of linear erosion is reduced. Runoff therefore takes place in washed hollows of uniform flatness, culminating in barely eroding rivers.

Inselbergs are characteristic elevated formations on peneplains. Selective chemical weathering of the surface can create jutting rock formations, sometimes of considerable height; intensive weathering can sharpen their edges, but the cores remain protected. Over time, erosion of the surrounding surface increases the relative height of the inselbergs.

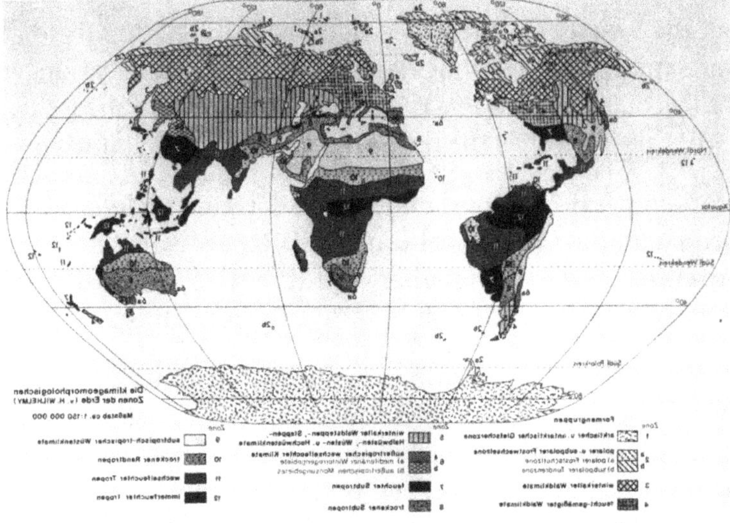

Figure.:The morphoclimatic zones of the Earth

The morphoclimatic zones of the Earth. Groups of landforms: 1. The arctic and antarctic zone of glaciers. 2. The polar-subpolar zone of alternating frost: a) the polar zone of frost-shattered debris; b) the subpolar tundra zone. 3. Forest climates with cold winters. 4. Humid midlatitude forest climates. 5. Cold winter climates of forested steppes, steppes, semi-deserts, deserts and desert highlands. 6. Extratropical climates of alternating humidity: a) mediterranean winter rain regions ; b) the extratropical monsoon region. 7. Humid subtropics. 8. Arid subtropics. 9. Subtropical – tropical desert climates. 10. Arid paratropics. 11. Tropics of alternating humidity. 12. Tropics of continual humidity.

Meteorological Threshold Values

The question in this case is to determine the extent in the direction of the pole of deep chemical weathering as the dominant morphodynamic process. In general, this extent is associated with the distribution of latosols. Their formation relies on chemical weathering, which is itself dependent on the combined factors of temperature and precipitation; since maximum weathering occurs at high temperatures with high precipitation, we need to find their minimum threshold values.

In German literature, these are normally defined by means of KOPPEN's climatic classification. It is assumed that the distribution of these soils can be correlated with the distribution of Kôppen's A-climates in most cases, and thus his aridity index and the 18 °C-isotherm for the coldest month are employed as criteria.

Taking precipitation as the primary criterion and subordinating temperature of the coldest month for the moment, the minimum threshold value is given as 360 mm (P = 2(18) [in cm]) for the hypothetical case in which the temperature of the coldest month is also the annual mean temperature in a region of winter rain. Where there is no marked seasonality, the threshold value is at least 500 mm (P = 2(18 +7)), and with sum mer rain at least 640 mm (P = 2(18 + 14))[2] .

It is better to present such an absolute threshold value, since chemical weathering results from the action of infiltrating water, and evaporating water is thus of no significance. The seasonal distribution of precipitation has also little value in this case.

KREBS has found that the isohyet of 500 mm is significant in relating the distribution of tropical inselbergs to annual precipitation. This coincides with PEDRO'S delineation, which is based on the empirical regional studies of other authors. The absence of more detailed quantitative data necessitates the acceptance of this isohyet as the threshold value for precipitation.

The threshold value for temperature comes under consideration in defining the limits of the tropical zone where it borders on the subtropical zone. PEDRO gives the annual isotherm of 15 °C as a suitable limit. In order to strike a medium between Kôppen's index (the 18 °C-isotherm for the coldest month) and Pedro's (annual isotherm of 15 °C), the annual isotherm of 18 °C is applied.

This combination of threshold values of 500 mm annual precipitation and 18 °C annual mean temperature is doubtless the assumption most open to question in a study of this kind. Pedro points out the discrepancy between the areal distribution of the soils under examination and of the processes responsible for their development. Ten years after publication of his treatise, he himself regards the given threshold value of precipitation as being too low.

However, when a value much higher than 500 mm is applied, the zone's limits become expanded to such an extent that its area becomes incongruent with the area of operation of its morphodynamic system. Therefore, for lack of a more appropriate formula, the combined threshold values of an annual precipitation of 500 mm and an annual mean temperature of 18 °C are used to define the limits of the tropical zone.

6. The Innertropical Zone of Landsliding and Valleys

Areal Distribution

This zone includes tropical rainforest regions of Central America; the northwest of South America, the Amazon region, the Congo Basin, parts of the northeast coast of the Gulf of Guinea, eastern parts of Madagascar, parts of Sri Lanka, the Malay Peninsula, major portions of the islands in southeast Asia.

Characteristic Landforms and Processes

Since temperatures and precipitation rates are even higher in these areas than in the tropical zone, chemical weathering here reaches maximum efficiency; nevertheless, landforms are very different.

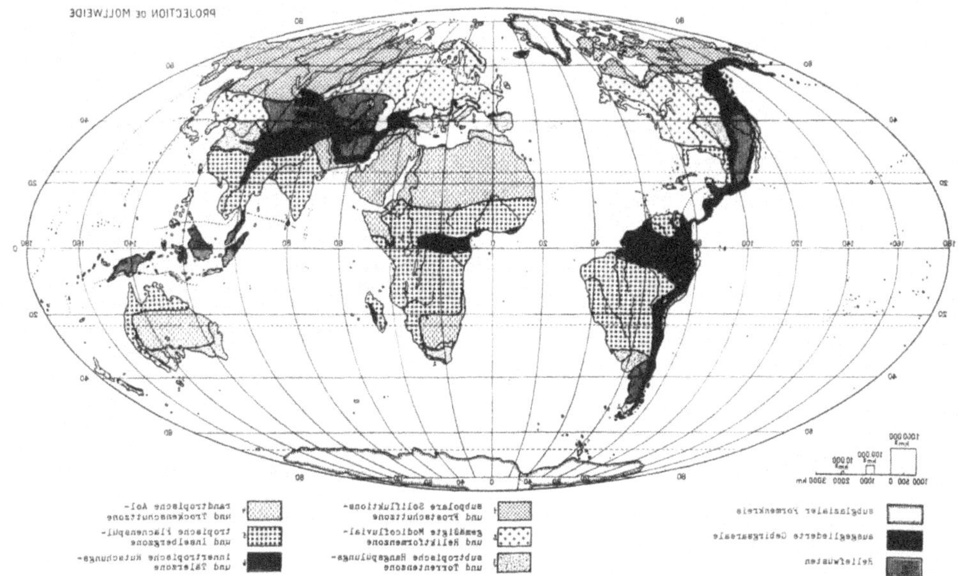

Figure: Climatic classification for geomorphological purposes

Climatic classification for geomorphological purposes. 1. The subpolar zone of solifluction and frost-shattered debris; 2. The moderate zone of modicofluvial action and relict forms. 3. The subtropical zone of slope wash and seasonal rivers. 4. The paratropical zone of eolian action and dry debris. 5. The tropical zone of sheet wash and inselbergs. 6. The innertropical zone of landsliding and valleys.

A striking feature is the sharp-edged flanks of mountains, which illustrate the high mobility of the soil in the form of niches. The numerous landslides are caused by a combination of several factors:

highly-efficient chemical weathering produces a thick covering layer of soil with great plasticity; high precipitation leads to saturation of the soil, thus diminishing friction; and frequent thunderstorms often trigger landslides.

Gravitational mass wasting is not limited to this one violent form. Continuous subsurface wasting also takes place where there are ground cavities produced by burrowing animals or root decomposition, and the results show as concave landforms. Their drainage also generally occurs through subsurface cavities.

Fluvial processes are of greater importance here than in the tropical zone. Lateral erosion is inhibited by dense vegetation, but abundant rainfall and highlyweathered soils provide the necessary conditions for the rapid erosion of river bottoms.

Meteorological Threshold Values

This survey of landforms and processes points out not only gradual but basic differences between the innertropical zone and the adjacent zone. Conditions for land formation seem to be similar to those of the tropical zone; for example, there is a thick covering layer of soil produced by chemical weathering in both zones. In order to define the limits, in light of such common traits, the critical factor of difference must be determined.

A study of the distribution of this zone in relation to tropical rainforest distribution reveals a high correlation which is not only areal but causal in nature. The vegetation of the rainforest protects the soil from direct climatic effects and is therefore a significant factor in the development of a morphodynamic system. Dense vegetation inhibits sheet erosion and promotes channel erosion; however, the absence of continuous spatial erosion contrasts with the effects of landslides and typical subsurface mass wasting.

The required meteorological threshold value should therefore be determined in light of the conditions responsible for the growth of the tropical rainforest. Particular attention is given to the level of precipitation, since this last also has a direct effect on landsliding and fluvial erosion. Because temperature is of negligible geomorphological consequence in the inner tropics, a definition based on this factor alone would be of little use.

In the literature, combination of minimum annual precipitation and a minimum number of humid months is unanimously accepted, although the values themselves differ. Since all recent studies suggest 1500 mm as the figure for minimum precipitation, it was not difficult to select an appropriate isohyet. The number of humid months has been determined by comparing the areal distribution of tropical rainforest with corresponding climatic diagrams. A minimum period of three months for duration of the non-humid season has proven appropriate, since any shorter interval gives misleading results, particularly in regions where annual precipitation is considerably higher than 1500 mm.

An additional threshold value of minimum temperature necessary for rainforest growth is not useful to geomorphological studies.

Therefore the annual isohyet of 1500 mm, together with an isohygromene of 9, are sufficient to define the limits of the innertropical zone.

Hydrogeomorphology

The term 'Hydrogeomorphology' designates the study of landforms as caused by the action of water. Water is one of the most important agent in forming and shaping of land forms.

Fundamentals of Hydrogeomorphology

The important fundamentals of hydrogeomorphological studies are given below:

- Mechanism and Process: Mechanism gives the explanation by describing the physicaland chemical effects while the process is the simultaneous operation of a set of specific mechanisms over a specific period of time. In hydrogeomorphological studies the landforms are studied with reference to groundwater conditions of the area. So while studying the mechanism and process in Hydrogeomorphology, the morphological, climatic and hydrological criteria's are considered.The study of mechanism and the process involved helps to give explanation behind morphology and distribution of landforms.
- Basictools required: To carry out a study in hydrogeomorphologyit is very important to first design a geographic database including both spatial and non spatial data requirements with their sources. The most important requirement for the studies in hydrogeomorphologyis maps. This includes the topographical maps, geological maps of the area, soil map, rainfall and climate distribution map, geomorphological map, population density map and groundwater fluctuation map. In India, the sources for these maps are various government agencies and publication like:

    1. Survey of india for toposheet maps.
    2. National bureau of census for population density map.
    3. Geological survey of india for geological maps.
    4. Indian meteorological division for climate and rainfall distribution datamaps.
    5. National bureau of soil survey and landuse planning, for soil maps.
    6. All india soil and land use bureau for landuse and land cover maps.

Along with the maps it also important to update the data through satellite data collected from various national and international remote sensing agencies like Indian institute of Remote Sensing Agency (IIRS), Dehradoon, National Remote Sensing Agency (NRSA), Hyderabad.

Properties of earth materials: Hydrogeomorpholgists studies the properties of earth materials to get better understanding of mechanism and process behind the geomorphic features. They emphasis on the properties of the parent material influencingthe formation and development of landforms along withthe hydrological conditions. Type of rock, weathered material, soil, superficial deposits, shear strength, porosity and mineral composition are some of the important properties that hydrogeomorphologists studies during their research.

Spatial scales of hydrology: Hydrogeomorphology not only studies the effect of hydrological processes on the geological processes but also analyses the effect of landforms on the hydrology of an area. Thus the study of scaling effects becomes essential. The spatial scales in hydrology can be distinguishedinto three types:

- Local scale: On local scale the parameters like slope angle directly influences the water flow path geometries, flow velocity and quantity.
- Hill slope scale: In hill slope scale is characterised by its runoff production. The soil properties influence the runoff production in the hill slope scale.
- Catchment scale: In catchment scale morphometry of the basin influences the runoff production.

Dimensions of hydrological units: Hydrogeomorphology studies the role of dimension of hydrological unitsin defining the size and dimensions of the geological landforms. In India Watersheds are delineated at various levels on the basis of the drainage network. Watershed can be defined as an area of land that includes common setof streams and river draining their water into single larger water body. Water in watershed can come from any of the sources of water from precipitation, in form of rain or snow or as groundwater or surface runoff. The size of the watershed depends on size of the stream, river, the point of interception of stream or river, the drainage density and its distribution.

Hydrological Cycle and Water Budget

Hydrology being an important part of hydrogeomorphology it becomes important for hydrogeomorpholgist to study the hydrological cycle, water balance and water budget for their research studies. Circulation of waterinto different forms between various spheresof earth surface is called hydrological cycle. Hydrological cycle recycles earth's water. The hydrological cycle includes inflow, outflow and storage of water at different levels. The inflows add the water into the system while outflow subtracts the water. Storage helps in the retention of water in the system.

Water budget studies the water availability. It includes the balance between the inflows and outflows of the water. In water or hydrologic budgetthe inputs are derived from precipitation, surface water inflow and groundwater inflow. Output factors includeevaporation, combined surface and groundwater outflow and transpiration.

Application of Remote Sensing and Geographic Information Systemsin Hydrogeomorphology

To get the proper understanding of hydrogeomorphology, it is essential to collect geological, structural and hydrological data of region. Collection of hydrological data i.e. assessing the data of surface and sub-surface water resources requires huge time and manpower. GIS and remote sensingare the platforms through which the all the required data can be collected with better accuracy and time. Remote Sensing and GIS carries great importance in hydrogeomorphological studies.

Before looking at the application of remote sensing and Geographic Information System (GIS) in hydrogeomorphology it is important to understand the concept of remote sensing and GIS.

In early 1960s the term remote sensing was first used for the first time. Remote sensing can be definedas the process of acquiring information through recording devices called sensors about the objects and phenomena without getting in physical contact with them. In remote sensing the data acquisition is done throughdifferent stages.

The stages involved in remote sensing for data collection are given below:

- Source of energy.
- Transmission of energy from the source to the earth surface.
- Interaction of energy with earth's surface features.
- Propagation of reflected energy/emitted energy through atmosphers.
- Detection of reflected /emitted energy by the sensor.
- Conversion of energy recorded into digital /photographic form.
- Extraction of the information from the data generated.
- Conversion of information into map/tabular form.

Satellite imageries and aerial photographs are the outputs of the process of remote sensing passing all the above mentioned stages. The interpretation of the satellite images and aerial photographs generated by remote sensing are made through the different elements of visual interpretations. The important elements are tone, texture, size, shape, shadow, pattern and association. In the studyof hydrogeomorphology the same elementsof visual interpretationare used to get theinformation through theinterpretation of the images generated by remote sensingof the landforms and the regions covering the water resources.

Geographic Information System (GIS) is a tool of digitally capturing, managing, analyzing and displaying geographic data using the computer hardware and software. Maps, computer hardware and software, information, procedures and people are the important components of GIS. GIS is an important tool for geographic studies because it analysis and answers real world problems. It makes dynamic maps and displays detailed information about the features in the maps. It not only displays but also studies and establish relationship between the features. The application of GIS can be seen in studying the relationships, patterns and trends of various spatial and non spatial elements of the earth surface. GIS contributes vastly in the study of Land use/land cover studies along with the studies associated with water and soil resources.

Over the years due to rapid increase in population, rapid urbanisation and industrialisation along with failure in monsoons has restricted the availability of surface water. Limited availability of surface water has increased the burden on groundwater resources. This has resulted into higher rates of groundwater withdrawal which further resulting in depletion of groundwater at alarming rates. For the management and conservation thestudy of groundwater resources is veryimportant. Satellite remote sensing has made it easy to study the spatial distribution of ground water prospects on the bases of geomorphology and other associated features.Satellite remote sensing is also a useful technique for groundwater exploration along with delineating the hydrogeomorphological units.Ground water accumulation, infiltration and movement largely depend on the factors like drainage, geomorphology, and slopeof the terrain, vegetation, soil and depth of weathering.All these factors can be easily studied using remote sensing at various levels.

Remote sensing not only studies the hydrological aspects but it also an effective tool for geological, structural, geomorphologic analysis and their mapping due toits synoptic, multi spectral and multi-temporal capabilities. Geologists largely depend on satellite imageries to collect the data on various lithological units.

The important advantagesof remote sensing in hydrogeomorphological studies are as follows:

- Remote sensing has access to large areas and even in inaccessible areas.
- The aerial photographs and satellite imageries provide detailed information about the uppermost layer of the earth surface which is essential for the hydrogeomorphological studies.
- Digital enhancement of satellite imageries improves the level of information useful in study of hydrogeomorphology.
- Data generated by remote sensing provide more accurate and spatial information in comparison tothe hydrogeological surveys.
- Through remote sensing hydrogeomorphological mapping of a terrain and analysis of their processes can be done easily.
- This will further help in soil resource mapping, groundwater potential zones demarcation, landscape ecological planning, hazard mapping and their environmental applications.
- The application of geomorphologic mapping using remote sensing can also be seen in land use planning and water resource management.

The IRS-1C and 1D data of WiFS, LISS-III and PAN sensors are highly useful for geological mapping. The WiFS cameragives the synoptic coverage of large areas and thus is useful for the regional scale mapping and understanding. The finer geological features like the traces of bedding and minor joints can be easily areidentified through panchromatic data. The panchromatic data provides detailed mapping while the multispectral LISS-III gives semi detailed mapping.

Along with remote sensing, in recent years several techniques have been developed in the field of Geographic Information System(GIS) throughwhichhydrogeomorphologicstudies can be conducted. In recent years GIS has emerged as a powerful tool in analysing the various aspects of groundwater occurrence. Groundwater resource of any area is largely controlled by the factors like lithology, structure, geomorphology, slope, drainageand landuse/land pattern. All these factors can be studied and analyzed as thematic layers using Geographic Information System (GIS).This makes it easy to delineate the groundwater prospect and deficit zones.

The main advantages of GIS in hydrogeomorphological studies are as follows:

- Large volume of data can analysed and integrated using GIS.
- Manipulates and analysis the individual layer of spatial data.
- Rapid, accurate and cost effective tool.
- GIS is a powerful tool for thegeneration of hydrogeomorphological mapping.

Application of Hydrogeomorphology

The application of hydrogeomorphologycan be seen in planning and management of various activities on the earth surface. Some of the important applications of hydrogeomorphology are as follows:

- The accurate, detailed, timely and reliable data on the extent, location and quality of land along with water resources and climatic characteristics helps resource planners in agricultural landuse.
- Data on land potential and conservation requirements through the hydrogeomorphological studies helps in improving the quality of land.
- Hydrogeomorphological studies are found helpfulby environmentalistsin identifying hazards and studying climate change.
- Geologist have found it useful in examining role of surface and subsurface flow regimes and flow paths on fluvial erosion and mass wasting.
- Ecologists found it useful to describe the linked water and geomorphic conditions that define habitats in wetlands,rivers and other environment.
- Watershed management depend on the collection and management of information on physical relationship between vegetation, soil and water resources, which can easily be done by hydrogeomorphological studies.

Future of Hydrogeomorphology as Geosciences

Over the years the relationship between physical and human environment is taking new turns and shape. The need for the new subfield called hydrogeomorphology stems from the emergence of various challengesresulted from growing human population combined with its effects on environmental and water resource systems. Transformation in hydrological cycle, water use, land use and climate is result of this change in man environment relationship. In today's era hydrogeomorphological studies are great option to manage and control the environmental problems. In recent years hydrogeomorphology has become an important branch of geosciences to tackle natural hazard impact, environmental auditing, resource assessment and impact assessment. To deal with the new challenges, hydrogeomorphology has developed close relationship withvarious fields like ecology, soil science and pedogeomorphology.

The growth in the importance of hydrogeomorpholgy as a geosciences can be seen with the fact that now the world organisations like UNDP and World Bank have added a clauseof understanding the hydrogeomorpholgy the area before starting any development project into it. Emphasis has been made by these organisations to have detailed knowledge of landforms, hydrogeology materials and earth surface processes to be utilized or the remedial work, planning framework and land zoning plans. Thus it can be said that the hydrogeomorphological knowledge is now being utilized in planning and development of earth as a whole. This shows the growth in status and responsibility of hydrogeomorpholgy in recent years as an important field of geosciences.

Applied Geomorphology

There has been an increasing recognition of the practical application of geomorphic principles and the findings of geomorphological research to human beings who are influenced by and, in turn, influence the surface features of the earth. Continuous increase in population has led to pressure on land resources, extension of agriculture to hilly and marginal lands resulted in man induced catastrophies like soil erosion, landslides, sedimentation and floods. A proper interpretation of landforms throws light upon the geologic history, structure, and lithology of a region. As geology becomes more specialized there is growing possibility that the application of geomorphology to problems of applied geology will be overlooked. The role of applied geomorphology relates mainly to the problems of analyzing and monitoring landscape forming processes that may arise from human interference. Human beings have over time tried to tame and modify geomorphic/environmental processes to suit their economic needs. Geomorphology has diverse application over a large area of human activity while geomorphologist may serve more effectively the need of society.

Geomorphology and Hydrology

Water either on the surface of the earth or groundwater used by human is available from different sources like streams, lakes and rivers. The lithological zones present different conditions of surface as well as groundwater.

Hydrology of Limestone Terrains

Comprehensive understanding of geomorphology is key to understand the hydrological problem of the limestone terrain. Limestone region yield more water than other due to its rock formation. Availability of water in limestone region depends on the type of rock. On the basis Permeability limestone rocks may be primary or secondary. Calcareous sediments decide the formation of rock and its primary permeability while earth movements in the form of tension and compression such as faulting, folding, warping, and due to solution or corrosion mechanism decide the secondary permeability.

Joints and fractures produced by diagenetic and diastrophic processes formed the secondary or acquired permeability which resulted into solution. The cavities of the solution in limestone region depends on whether it has been situated in the past or allow joints and bedding planes to be actively more enlarged. In Florida (USA), these solution cavities are common at considerable depth in the Tertiary limestones. The significance of solutional opening with increased permeability is important in present day topography but also in karst landscape too.

Geomorphology plays an important to obtained water in limestone region. It may be easy or difficult to obtain water from wells in a limestone terrain. There may not be difficulty in obtaining wells of large yields if the limestones have enough permeability and are capped with sandstone layer. In such case the yield of water may be low or inadequate, but subject to contamination. Karst plains lacks filtering cover, and any swallow holes, sinkholes, or karst valleys within an area of clastic rocks should cast doubt upon the purity of the water of springs.

Glaciated Areas and Groundwater

Preglacial and glacial time history, types of deposits and landforms determine the possibilities of large supplies of groundwater potentials in glaciated regions. Yield of large volume of water obtained from Outwash plains, valley trains, and intertill gravels or buried outwash. Due to clay content most of the aquifers are poor, but containing local strata of sand and gravel may hold and supply enough water for domestic needs. The study of preglacial topography and geomorphic history of the area could detect the presence and absence of underground water.

Geomorphology and Mineral Exploration

There is a close association of geological structure and minerals deposits. Characteristic of landscapes of specific areas could indicate these geological structures. Economic geologist has not appreciated the exploration of some minerals in the name of understanding of the geomorphic features and history of a region. In search for mineral deposits, these three points may serve for geomorphic features as:

1. Some mineral have direct topographic expression for its deposits;
2. The geologic structure and topography of an area have correlation which clue the accumulation of minerals;
3. Geomorphic history clearly indicates the physical condition under which the minerals accumulated or were enriched of a particular area.

Surface Expression of Ore Bodies

Some of ore bodies have surface expression, but many do as topographic forms, as outcrops of ore, gossan, or residual minerals, or as such structural features as faults, fractures, and breccia zones. It is not necessary that all ore outcrops are reflected in positive topographic forms. The lead-zinc lode could be marked by a conspicuous ridge in the case of Broken Hill, Australia. Quartz veins could stand out prominently as they are much more resistant to erosion than the unsilicified rocks, as in Chihuahua, Mexico. Some veins and mineralized areas may lack conspicuous topographic expression or it may be reflected by subsidence features or depressions. Though no generalization can be made about the exact type of topography necessary for the iron ore accumulation, distinct topographic expression is needed for a particular deposit. Residual iron deposits are the results of concentration of iron due to long periods of weathering, and thus for their accumulation, old erosion and weathering surfaces are favorable sites.

Weathering Residues

Geomorphology can play an important role for several important economic minerals which are essentially weathering residues of present or ancient geomorphic cycles. Apart from iron deposits, materials like clay minerals, caliche, bauxite and some manganese and nickel ores are of this nature. Recent weathering surfaces may exhibit residual weathering products or it may lie upon ancient weathering surfaces which are now buried.

Peneplain or near peneplain surfaces are most commonly surfaces upon which they form. In general such minerals are to be found upon remnants of tertiary erosional surfaces above present base levels of erosion. It is not yet clear why the weathering of igneous rocks produces both clay minerals or hydrous aluminum silicates and hydrous oxides of aluminum, such as bauxite. The difference in the final product is determined by the climatic conditions under which weathering takes could be one of the explanation.

The residual products from the weathering of igneous rocks are clay minerals found in temperate climates known as kaolinization. It should be recognized that numerous minerals other than kaolin may form in same climate. On the contrary, under tropical climates final weathering products are hydrous oxides of such metals as aluminum, manganese and iron. This type of weathering is known as laterization. The phase of geology which concerns with the recognition and the study of ancient weathering surfaces and soil has come to be known as paleopedology. Though it offers many possibilities but still in its infancy in the search for the type of mineral deposits designated as weathering residues of geological phase.

Epigenetic Minerals and Unconformities

Ancient erosion surfaces are associated with numerous deposits of Epigenetic minerals. Mills and Eyrich emphasized the role played by unconformities in the localization of mineral deposits. The mineral deposits found from the ranging age of Precambrian to Tertiary, shown evidences of close association with unconformities in districts of US and Canada, such minerals are uranium, vanadium, copper, barite, fluorite, lead, nickel, and manganese. There is constant work of weathering and erosion on the rocks of earth's surface and this weathering work has economical value of rock product.

Placer Deposits

Placer deposits are mixtures of heavy metals with specific location, geomorphic principles have been applied other than any other phase of economic geology. Geomorphic processes are the main cause of placer concentration of minerals, found in specific positions with distinctive topographic expression. the deposition of placers affected by the type of rock forming the bedrock floor. There are as many as nine types of placer deposits. They are residual or 'seam diggings', colluvial, eolian, bajada, beach, glacial including those in end moraines and valley trains, and buried and ancient placers. The most important among them is alluvial placers.

The other name of residual placers is 'seam diggings' which are residues from the weathering of quartz stringers or veins, are usually of partial amount, and grade down into lodes. Creep down slope is the main reason for the production of colluvial placers and are thus transitional between residual placers and alluvial placers.

Most of the gold placers of this form have been found in California, Australia, New Zealand, and elsewhere. Colluvial placers (the koelits) and alluvial placers (the kaksas) are parts of the tin placers of Malaya. . The most important minerals like gold, tin and diamonds are obtained from alluvial placers. South Africa's diamonds from Vaal and Orange River districts, the Lichtenburg area, the Belgian Congo, and Brazil's Minas Geraes, are obtained from alluvial placers. Placer deposits have total share of around 20 per cent of world's diamonds. Australia, lower California, and Mexico have yielded gold in aeolian placers. Gold in California and Alaska, diamonds in the Namaqualand district of South Africa, zircon in India, Brazil, and Australia, and ilmenite and monazite from Travancore, India have yielded from beach placers.

Gold placer

Oil Exploration

Several oil fields have been discovered because of their striking topographic expression. These oil fields are characterized by anticlinal structures which strikingly reflected in the topography. When viewed from aerial photographs, many of the Gulf Coast salt dome structures are evident in the topography. For the student of geomorphology, it is fairly good working principle to suspect that areas that are topographically high may also be structurally high, where possibilities of topographical inversion at the crest of a structural high may result with weak beds.

In regions of heavy tropical forest, topography cannot be seen through the intense forest cover, an anticlinal or domal structure may outline due to the tonal differences in the vegetation. In search for oil, more subtle evidence of geologic structures favourable to oil accumulation is being made. Aerial photography is one such technique through which drainage analysis of a terrain can be shown. Drainage analysis is useful particularly in regions where rocks have low dips and the topographic relief is slight. Permeability may be either primary or secondary in carbonate rocks. Number of large oil yields from limestone has been obtained from rocks which have a high degree of permeability produced by solution.

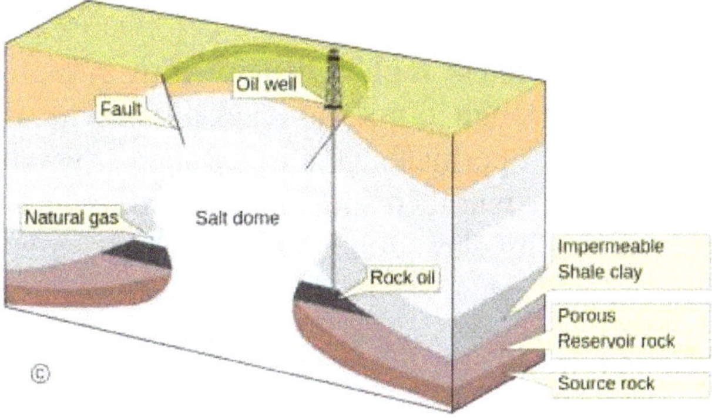

Salt dome

Elongate buried sand bodies are basically shoestring sands. Probably there is no phase of petroleum exploration which can use to better advantage a knowledge of the in depth characteristics of specific topographic features than that which deals with the misuse of shoestring sands. Most of the oil and gas sources are associated with unconformities - ancient erosion surfaces; hence a petroleum geologist must deal with buried landscapes.

Geomorphology and Engineering Works

Evaluation of geologic factors of one type or another often involve in most of the engineering projects, among all the factors terrain characteristics is most common. A detailed study of the geomorphic history of an area may support the proper evaluation of surficial materials and the bedrock profile configuration.

Road Construction

Topographic features of an area determined the most feasible highway route. Road engineering faces a number of problems by different types of terrain that includes geologic structure, geomorphic history of the area, lithological and stratigraphic characteristics and strength of the surficial deposits. Area like karst plain required repeated cut and fill, if not done then the road will be flooded after heavy rains with surface runoff from the sinkholes.

The presence of enlarged solutional cavities in karst region emphasis on the designed of roads in such a way that road should not be weakened. Region like glacial terrain presents a number of engineering problems. Road construction in flat till plain is topographically ideal but other areas where moraines, eksers, kames or drumlins like features exist there is need for cut and fill to avoid circuitous routes. Areas which are characterized by late, youth and maturaity of relief will require more bridge construction and many cuts and fills. These types of areas are consistently facing problems like landslides, earth flows, and slumping.

Landslides and different types of mass-wasting present problems not only in different phases of engineering but in highway construction also. Subgrade or the soil beneath a road surface has become more significant because of its control over the drainage beneath a highway, therefore construction design of highway should be in such a way to carry heavy traffic. Two factors largely determine the lifetime of a highway under moderate loads is the quality of the aggregate used in the highway and the soil texture and subgrade drainage. The type of parent material and the relationships of soils to its varying topographic conditions are more essential in modern road construction.

The most serious problems encountered by highway engineers is pumping which means expulsion of water from beneath road slabs through joints and cracks. It is evident that pumping is particularly greater over glacial till than over permeable materials such as wind-blown sand and outwash gravel. Poor drainage in a subgrade is mainly responsible for pumping. Poor and best performance of the highway is characterized by silty-clay subgrades with a high water table and granular materials with a low water table respectively.

Dam Site Selection

A synthesis of knowledge concerning the geomorphology, lithology, and geologic structure of terrains has greatly helped while selecting sites for dam construction. According to Bryan, five main requirements of good reservoir sites depend on geologic conditions:

(1) Adequate size water-tight basin;

(2) A narrow outlet of the basin with a foundation that will permit economical construction of a dam;

(3) To build an adequate and safe spillway to carry excess waters;

(4) Availability of resources needed for dam construction (earthen dams); and

(5) Assurance that excessive deposition of mud and silt will not short the life of reservoir.

Constructing a dam in a limestone terrain may prove a difficult one, for instance, the Hondo reservoir was built over limestone in southeastern New Mexico with a water table some 20 feet below the surface. Rapid leakage was the cause to abandonment of the reservoir. Building a dam in a valley may not be a good dam site from the standpoint of the size of the dam. Buried bedrock valleys containing sand and gravel fills are common in glaciated areas, which may not depict adequate picture of surface condition. Making dam on those sites where subsurface topography is not supportive with buried preglacial valley with sand and gravel in it would have a chance of leakage.

Hondo reservoir in New Mexico

Location of Sand and Gravel Pits

Sand and gravel have more commercial and industrial uses than many engineering. Evaluation of geologic factors such as variation in grade sizes, lithologic composition, degree of weathering, amount of overburden, and continuity of the deposits are important while selecting suitable sites for sand and gravel pits. Floodplain, river terrace, alluvial fan and cone, talus, wind-blown, residual, and glacial deposits of various types are areas where sand and gravel may be found in abundance. In recent years, there is a great demand of gravel than sand due to decreased use of plaster in home construction therefore knowledge of various grade sizes is more important.

Sand and gravel pits

There are high proportions of silt and sand in floodplain deposits which show many variable and vertical gradations and heterogeneous lateral. With their angular shape as well as variable in size alluvial fan and cone gravels are found near their apices. Being angular like talus materials are too large to be useful and are limited in extent. There is only sand in wind-blown sands but have no gravel. Residual deposits are likely to contain pebbles that are suitable for cement work. These residuals are also limited in extent. Favorable sites for pits are terraced valley trains and outwash plains, which are usually extensive and do not have a thick overburden. Due to its large amount of the material, kame deposits show a poor degree of assortment because it discarded on the ground of too large or too fine.

Geomorphology and Military Geology

Allied powers during world war were slow to make the maximum use of geology in warfare. Geologist were utilized but to a limited extent in World War I. Before military authorities saw the needs for and possibilities of the use of geologic experts, the war was well-advanced. During wars the information that was useful was more geologic than geomorphic in nature. The information regarding digging trenches, mining, countermining, and water supply or other material was not utilized. Topography became more important during World War II with the development of the blitzkrieg type of warfare, because effectiveness of a blitz depends to a large extent upon the trafficability of the terrain. In recent years terrain appreciation or terrain analysis have become more important with military.

Trenches during World War

For a terrain if geological maps fail somewhere, geological principal can be applied with advantage to interpreting the terrain from aerial photographs. Little training required to recognize features like mountains, hills, lakes, rivers, woods, plains or some kinds of swamps. It is important to know the kind of hill, plain, river or lake, and so on, because by knowing this it is quite possible to reconstruct the geology of that region. Aerial photographs are useful for the preparation of terrain intelligence as they provide information on the geology of the area. Terrain has been an important factor in the Korean War and in the fighting in Vietnam region. With the development of atomic bomb and ballistic missiles, topography would no longer play an important role in wars but its confine to the local areas for war purpose.

Geomorphology and Regional Planning

Geomorphologic information can be utilized at various levels of planning. Combination of topographic information, soils, hydrology, lithology, terrain characteristics and engineering included on terrain maps make suitable for regional planning. Applied geomorphology has distinct place in regional planning. At broadest scale it can be used as delineate areas for forest, mountain, plateau, recreational, rural and urban areas. A balanced growth of a country's economy requires a careful understanding of its natural resources and human resources. Rural or underdeveloped terrain fulfills a variety of recreational needs. There is a transformation from a terrain maps into land-use suitability maps to develop rural and urban areas. Detailed information on topography enlightened regional planners who may then advise development projects best suited for separate region.

Geomorphology and Urbanisation

There is a separate branch known as urban geomorphology applied to urban development. According to R.U. Cooke, this branch of geomorphology is concerned with "the study of landforms and their related processes, materials and hazards, ways that are beneficial to planning, development and management of urbanized areas where urban growth is expected".

Geomorphic features decide the stability, safety, basic needs and even its expansion. That means city or towns entirely depends on lithological and topographical features, hydrological conditions and geomorphic features. Urban geomorphologist commence even before urban development through field survey, terrain classification, identification and selection of alternative sites for settlements irrespective of plain or hilly areas. These urban geomorphologists would be concerned with impact of natural events on the urban community and that of urban development on the environment.

Urban morphology

When geomorphological problems not understood by the planners and engineers then it leads to destruction and damage to urban settlements in different environmental regions. Settling of foundation material in dry or glacial region, weathering process, damages of roads and buildings through floods in many parts of the world are not a recent phenomenon. These problems arise due to misunderstanding of the geomorphological conditions. In developing countries attention has not been given to the geomorphological conditions before the development of existing urban centres. This leads to haphazard growth of city with squatter settlement and shanty towns with urban morphology.

Geomorphology and Coastal Zone Management

Coastal zones are not in linear as a boundary between land and water rather viewed as dynamic region of interface of land and water. The major threat to the fragile coastal zone is its deteriorating coastal environment through shoreline erosion, loss of natural beauty, pollution and extinction of species coastal zone management requires an integrated approach. The most widespread material is beach sand, found mainly in low latitudes. Beach sand and gravel is widely used for construction industry.

Breakwater

Geomorphologists have made some significant contribution towards an understanding of shoreline equilibrium in Eastern Australia where it considerable development of sand mining for heavy minerals has been done. Some measures have been designed or coast protection includes sea-defence structures such as seawalls, breakwaters, jetties and groynes. To protect the sea backshore zone from direct erosion cut, sea walls are designed since these walls are impermeable they increase the backwash and produce a destructive wave effect. Breakwaters can be built either normal or parallel to the coast. It is necessary to monitor and quantify wave conditions, tidal currents and sediment movement in the nearshore zone to evaluate how sea defenses and other man-made structures affect shoreline equilibrium.

In context of coastal zone management Hails emphasizes that applied geomorphology must be concerned with quantitative and not descriptive research in order to obtain relevant and accurate data on (i) natural erosion and deposition rate (ii) at what rates and amount the sediment transport from river catchments to the near shore zone; (iii) variations in sediment composition and offshore distribution; (iv) sand supply sources and shoreline equilibrium; (v) interchange rate of sand between beaches and dune systems; (vi) the effects of constructing sea defences; (vii) offshore sediment dispersal and the dredging effects of seabed morphology, sediment transport and wave refraction; and (viii) analysis of landform including topography of the nearshore zone, form of the continental shelf and of relict coast lines, particularly in terms of rock outcrops. Above investigation provides relevant baseline data needed for systematic planning process and monitoring programmes but also for land use scheme.

Coastal erosion

Geomorphology and Hazard Management

Hazards can be put in natural or man-induced where tolerable level or unexpected nature exceeds. According to Chorley, geomorphic hazard may be defined as "any change, natural or man- made, that may affect the geomorphic stability of a landform to the adversity of living things". These hazards may arise from immediate and sudden movements like volcanic eruptions, earthquakes, landslides, avalanches, floods, etc. Faulting, folding, warping, uplifting, subsidence, or vegetation changes and hydrologic regime due to climatic change arise from the long term factors. Areas having past case histories of volcanism and seismic events help in making predictions of possible eruptions and earthquakes respectively. Regular monitoring of seismic waves, measurement of temperature of craters lake, hot springs, geysers and changes in the configuration of volcanoes whether dormant or extinct can reduce the hazard to some extent. A detailed knowledge of topography can predict the path of lava flow and its eruptions points in advance

Path of lava flow

The behavior of a river system can be well understood by its geomorphic knowledge through its channel, morphology, flow pattern, river metamorphosis and so on. It may help controlling excess water in river and control measures during flood season. Prior knowledge of erosion in the upper catchment area and carrying sediments to its proportion may help in understand the gradual rise in river bed, which may lead to levee breached and cause sudden floods. Earthquakes may be man induced or natural geomorphic hazards. Detailed study of seismic waves region would help in identifying and mapping the zones of high to low intensity to reduce the risk of human life.

Other Applications of Geomorphology

Some of the applications of geomorphic principles have been used in applied geomorphology but there are other fields where geomorphic knowledge of terrain is more important. Soils maps to some extent are topographic maps and difference in soil series fundamentally rest upon topographic conditions under which each portion of soil series developed. Soil erosion related problem is essentially a problem involving recognition and proper control of such geomorphic processes like sheet wash erosion, gulleying, mass-wasting, and stream erosion. The angle of slope is not a single factor determined the severity of erosion.

With the introduction of air photographs and satellite imageries preparation of specialized maps and interpreting them has become easier and more accurate. Now a days, aerial photographs are being used for evaluating landforms and land use for city developmental plans, construction projects, highway etc. Another tool i.e. Remote sensing is necessary for sustainable management of natural resources like soil, forest, crops, oceans, urban and town planning etc. At present Geographical Information Systems (GIS) technology has been used along with Remote Sensing techniques in geomorphic features interpretation.

All fields discussed in this chapter should be sufficient to show an understanding of geomorphic principal, besides the geomorphic history of a particular region, geomorphic features may contribute in applied geology to the solutions of problems. To control the adverse effects of human activities on geomorphic forms and processes, application of geomorphology can be of immense use.

Urban Geomorphology

The processes and forms involved in urban activity as an earth surface process are the subject of urban geomorphology. Urban geomorphology examines the geomorphic constraints on urban development and the suitability of different landforms for specific urban uses; the impacts of urban activities on earth surface processes, especially during construction; the landforms created by urbanization, including land reclamation and waste disposal; and the geomorphic consequences of the extractive industries in and around urban areas. The diversity of urban substrates is a consequence of their geomorphic history, the ways in which past environmental changes, including climatic and sea level changes have affected the form of land and the types of surface materials.

Urban geomorphology is a key element in supplying the guidance needed to achieve a better quality of urban life and working towards more sustainable use of resources.

Geomorphology has a valuable role to play in the planning, development, and management of urban areas in dry lands, especially in evaluating terrain prior to urban development, and in monitoring changes during and after development. The type of information required depends chiefly on the local geomorphology and the responsibilities and altitudes of the relevant agencies. It will also vary with the phase of regional planning, city planning, site planning and development, andpost construction management all require information that can be provided by the geomorphologists.

Much work in geomorphology is of great potential value to man in his use of the physical environment, and theapplication of geomorphologic knowledge has increased in recent years, in harmony with growing public and political awareness of environmental problems and specifically because geomorphologists have come to give readers attention to those aspects of the subject of greatest practical application-the dynamic relations between landforms, materials, andcontemporary processes. As far as Applied Urban Geomorphology is concerned it is the study of landforms, and their related processes, materials, and hazards, in ways that are beneficial to planning, development, and management of urbanized areas or areas where urban growth is expected.

Geomorphic knowledge is of great significance in deciding the urban growth and urban morphometryin geomorphological fragile zone. Since urbanization is continuesprocess of urban growth, it is being affected by several factors. An urban agglomeration denotes a continues urban spread and normally consists of a town in its adjoining urban outgrowth. Geomorphic aspect plays a great role in deciding the organizational implications.

One important qualification must be made about urban geomorphology: different aspects of geomorphology have been studied in urban areas for many years-not only by geomorphologists, but also by engineers and others, some of whom may neverhave heard of the science. It is not, and never has been the exclusive preserve of the geomorphologist. For example, engineers in Los Angeles have studied the movement of sediment into, through, and out of the metropolis fordecades, and the informationthey have collected has been used to predict, inter alia, the life span of reservoirs. Nevertheless, as the field of urban geomorphology is one in which many different aspects of the environment are closely related and can be beneficially integrated, and is also one of rapidly growing knowledge requiring greater specialization among those studying it, it is scarcely surprising that planners and engineers are increasingly either receiving geomorphological training themselves or turning to specialist geomorphologists for help in tackling problems of a geomorphological nature.

Recent study suggests that there will be significant variations in the amount of urban expansion: most of the urban expansion will occur in China; some specific regions will have high probability of urban expansion; some regions will have low probability of urban growth.

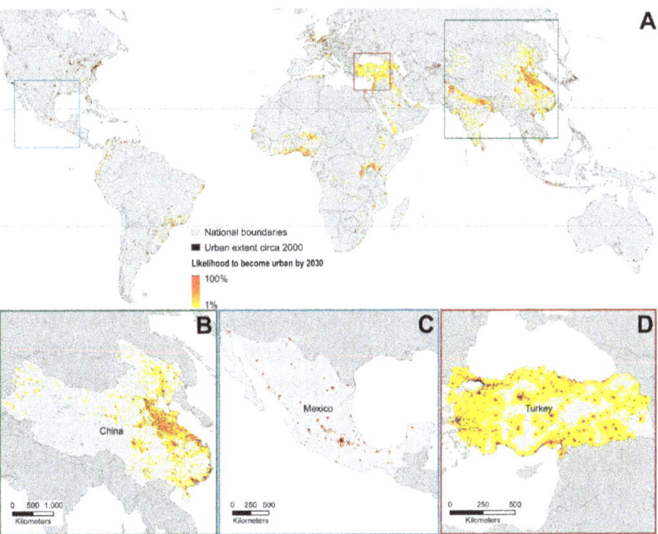

Figure: Urbanexpansion probabilities at global level

A Bibliographic Review in Urban Geomorphology

Wolman was among the first geomorphologists to measure the physical impacts of urbanization on watersheds and stream channels. His studies found that average sediment production rates are moderate to high during pre-urbanagricultural uses of land, followed by a spike during construction,and finally a decrease in sediment yield after urbanization. Wolman's results indicate that the channel response following urbanization is a period of deepening followed by lateral migration and channel widening. These observations of the changes in channel form are consistent with other work carried out during the same time period when fluvial geomorphology in the urban environment was a developing scene.

Graf observed the effects of rapid urbanization on the fluvial geomorphology of two small watersheds near Denver, Colorado. Graf found that the initialimpact for these sites was extreme aggradationand increased flood plain access due to increased rates of upslopeerosion. The secondary impact, after the watershed development was nearly complete and impervious cover in place, was vertical incision and down cuttingthrough the previously aggraded material. Arnold, C.L., P.J. Boison, and P.C. Patton, conducted a similar geomorphic and hydrologic study of a small urbanizing watershed. The frequencyof bank full discharge increased, and changes in the sediment regime were consistent with Wolman's observationdescribing the effect of urbanization. Elsewhere, in other parts of the world, studies in urban geomorphology have gained ground. Notable studies on this line worth mentioning are "geomorpohology and urban development in Manchester area"by Ian Douglas. This work underlined the impact of geomorphology on river dynamics, urban growth, glacial deposits, subsidence, sewer collapse and ground conditions. There are also other works such as urban geomorphology in Dry Lands.

This study was undertaken as a consequence of serious soil erosion, landslides and widespread flooding where hundreds of people were killed and thousandsof homes ruined. The dominant environment processes responsible for this crisis are geomorphological problems, problems relating tothe nature of land surface and the forces that act upon it. In Indiathe studyof urban geomorphology, first appearedin 1988. This study was experiencedin Mussoorie and its Environs' by H. Prasad.This study underlined the impact of geomorphology in identifying areas for establishment of new settlements.

Attention to urban geomorphologyhas increased in recent years in harmony with the growing recognition of the importance of the much broader, but closely related, fields of environmental and urban geology. Although some aspects of urban geomorphology have been considered in recent books, such as those by Coates, Detwyler and Marcus, Legget, Cooke and Doornkamp, and Leveson, the rationale of the subject has not been clearly formulated.

Urban geomorphology combines the ambient geology, landforms, and geomorphological processes with the evaluation of impacts brought to these by urbanization. The practitioners of urban geomorphology tend to concentrate on alteration, using the ambient physical environment as a baseline. A number of case studies from different parts of the world (dealing with topics such as slope instability, seismic hazards, increased flood problems, and land subsidence) have demonstrated the utility of urban geomorphology to engineers, city managers, and urban planner.

Objectives of Urban Geomorphology

Objectives of Geomorphological Appraisal Prior to Urban Development

The urban planner's primary environmental requirement prior to urban development is a knowledge of the nature and disposition of natural resources and hazards. The principal objectives, therefore, of surveys designed to satisfy this requirement are the identification of the range of possible locations of resources and hazards and to analyse conditions within suitablelocations, to use the environmental resources more economically, beneficially, and efficiently.

Within these broad objectives, the geomorphologists commonly haveseveral aims: (1) to preventurban growth from destroying valuable resources; (2) to identify and evaluateland and material resources required o development; (3) to limitundesirable impact of urban development on geomorphological conditions; (4) to predictthe potential responses of ground surfaces to urban development; and (5) to assessthe potential impact of geomorphological hazards on the urban community.

Geomorphologists commonly adopt one or more of the three major approaches to appraisal prior to urban development. First, by far the most profitable and widelyused approach is that of formally classifying and describing terrain features, through morphological or geomorphological mapping and/or the interpretation of air photographs or other remote sensing imagery. Second, analysis of process dynamics and landform change may be accomplished through, for example, the analysis of historical records (e.g. climatic and hydrological data of a region). A third approachis to appraise one, poorly known situation by analogy with another similar but better-documented situation elsewhere. This approachis, of course, dependent on the availability of data from the analogous situations and it provides a strong argument for the collection of information in data banks such as that envisaged by the VIGIL network and in deserts by Bekett and others.

Before urban development, thegeomorphological contribution is mainly of importance in providing surveys of direct use in themselves, of value as a source of derivative maps etc., and as a basis for more detailed subsequent surveys.

Objectives of Geomorphological Appraisal During and after Urban Development

During and after urban development the urban planner normally require to know the effects of natural events and circumstances on the urban community, and the impact of urban development on the environment. It is to be noted that the primary interest ofthe planner is to understand the environmental consequences of urban growth. Within this primary objective, planning and management aims include (1) the minimization of environmental impact; (2) the development of local, spatial and temporal data bases from monitoring studies to formulate the urban development plans; and (3) the continues evaluationof plans, management organizations, and procedures to ensure an harmonius environmental management.

The main aim of geomorphological work in this context is to monitor the dynamics of geomorphological systems with a view to predicting spatial and temporal changes in a way that allows the planner to respond effectively and in good time. Geomorphological approaches to appraisal during and after urban development are similar to those adopted prior to development, but their relative importance changes. Field monitoring is pre-eminent, whether it is on a global scale or a city wide scale. Monitoring often requires the establishment of fixed observation stations.

Examples of the Value of Geomorphological Information in the Planning Process

Geomorphology in city planning: If, in the planning of a new city the geomorphological surveys are carried out before the settlement of the city, then we can avoid incongruity between the environmental conditions and the city. For example, suppose urban plannersmaking the city plan of a city, the first city plan (we can term it is as city plan A) has an attractive spatial geometry, but it would have encountered several problems-building over scarce aggregates; hydro-compaction; flooding, sedimentation and erosion or roads crossing alluvial-fan channels; blowing sand and salt weathering. In the revised city plan (we can term it as City plan B), which attempts to accommodate the implications of the geomorphological map, most of these problems are either avoided orsensibly controlled, thus saving resources, time and money.

At the scale of a whole city, a common problem is that the management of a single, natural unit, such as a drainage basin, is divided between several administrative organizations, with resulting-duplication or dispersion of effort, and perhaps competition and conflict. This isa problem that can be avoided if environmental criteria are used intelligently in the initial formulation of responsibilities of different authorities-although it would benaïve to assume that other factors are not usually more important in such formulations.

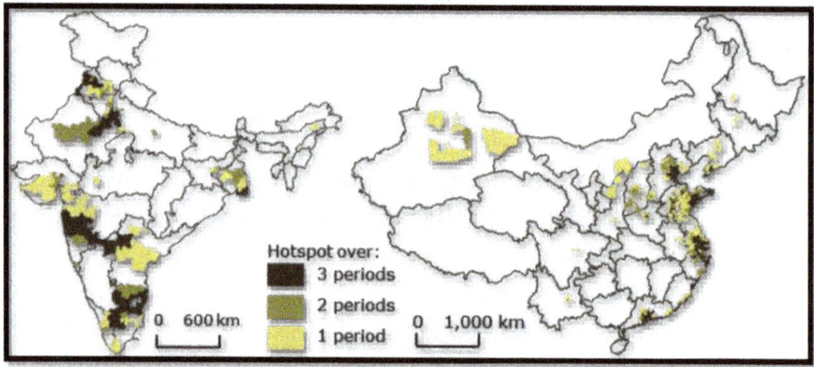

Figure: Urban cluster growth hotspots detected

Geomorphology and site planning and development: An excellent example of the way in which geomorphological advice can beneficially modify site development plans is provided by Mader and Crowder. He citedthe examples from USA's hilly country areas. In 1956 a residential development was proposed in the hill country of the growing settlement of Portola Valley, south of San Francisco, California. Subsequent geological and geomorphological studies related to the formulation of the town's general plan and zoning and subdivision ordinances revealed the nature and extent of potential slope instability in the proposed development area. A relative slope stability map showed areas of stable, potentially moving, andmoving ground, and located major landslides. That map, together with other relevant information formed the basis for a new plan, in which houses were clustered on stable ridge crests, and the number of lots was only slightly lower than the maximum permitted under the general formula relating lot sizeto average slope; even more important, about

15 houses sites and some roads on the original plan were removed from actively moving ground, and considerably more house sites were removed from potentially moving ground.

There are numerous instances where specific geomorphological information might have helped to avoid problems and made the site-development process more efficient.

Geomorphology and post-construction management: It is suggested that major environmental hazards can be avoided by monitoring of geomorphological processes and landform changes after construction and this can also help environmental managers and policy makers in influencing future policy planning. Ruby's study of sediment-yield trendsin the Los Angeles River catchment is a straightforward example. The accumulation of sediment in debris basins at the mouths of mountain canyons provides a rough measures of sediment yields and allows the performance of smaller check dams to be evaluated. Ruby compared accumulated sediment yields in one canyon (Dunsmore) before and after check-dam construction with the regional norm. There are many problems associated with the data and their interpretation, but regression lines relating sediment yield in Dunsmore canyonto the regional norm for 30 years of records show that for the first 21 years the canyon performed similarly to the regional watershed, that during the ten years following construction of check dams sediment yieldfrom the canyon relatively declined, and that the decline has become progressively less over time. In the period since check dam construction sediment yield was reduced overall by approximately half, but there now appears to be a trend back to pre-treatment yields, indicating a decline in the trap efficiency of the check dams. Clearly an alternative strategy for sediment control is required.

Before geomorphological problems of any urban development are reviewed systematically, two further introductory themes require examination: the availability of information relevant to geomorphological studies in urban development area, and the integration of geomorphological data into the whole assemblage of environmental data relevant to planning decisions.

Relations between Geomorphology and other Scientific Information of Value to Planners of Urban Areas

Geomorphological information forms but one part of the body of environmental data that may be of value to urban planners and engineers. Many environmental attributes of interest are both closely linked and highly interdependent. Data on the regional setting is primarily relevant in assessing hazards and resources and in choosing locations for development; data on site conditions related to decisions about specific developments. On the basis of recent studies it can be argued that geomorphological surveys can provide a useful first stage in the environmental assessment, not only because geomorphology is a fundamental basis for urban development, but also because studies of geology, soils, hydrology, etc., can all benefit from, and be facilitated by access to geomorphological surveys.

Geomorphology and Environmental Impact Assessment

On the basis of above discussion it is important to record that demands are increasing from planners for wide-ranging environmental reports prior to making and implementing planning decisions. The most important new requirement is not so much for resource survey but for studies to evaluate the impact a proposed development is likely to have on the physical environment.

Although many planning authorities have required environmental impact assessment for years, recent legislations has enforced and codified the requirement in several countries, providing a substantial stimulus to the development of methods for assessment and for integrating environmental data. The most important new law was NEPA (National Environmental Policy Act), 1969,passed by the US government,and this was followed by similar measures in other countries such as Australia and Israel.

Figure: Shenzhen CBD and Mai Po Marshes of Hong Kong.

To assess potential environmental impact is extraordinarily difficult because it involves predicting complex responseson the basis of what is usually woefully inadequate scientific data. However, political necessity has prompted several attempts to develop standardized techniques of assessment in order to streamline production of reports, to facilitate comparisons, and to simplify the preparation and presentation of complex problems.

Figure: The Chinese city of Sanghai will be one of the largest urban areas in the world.

Impact of environment on development: This theme has much in common with the previous topic and the techniques of environmental impact assessment mentioned there could well be appropriate here, for clearly cause and effect are intimately related in man's relations with his environment. But emphasis in this field commonly rests mainly on the assessment of natural hazards at different scales. Thus, a major problem of environmental planning at a regional scale is to establish priorities. In this context, assessment of the potential relative impact of environmental hazards is particularly important, and such impacts will vary both spatially and temporally.

Geomorphology has a valuable role to play in the planning, development and management of urban areas, especially in evaluating terrain prior to urban development, and monitoring changes-during and after urban development. Responsibility for studying and managing geomorphological-resources, hazards, and other problems may rest with a varietyof agencies within local hierarchies of management organizations. The type of information required will depend chiefly on the local geomorphology and the responsibilities and attitudes of the relevant agencies.It will also vary with the phase of planning: regional planning, city planning, site planning and development and post construction management all required information that can be provided by the geomorphologist. In such circumstances, he must invariably be able to use a wide range of relevant data sources, some ofwhich may be difficult to obtain. Commonly, the geomorphological information must be integrated into a broader assemblage of environmental information that is useful to planners: among the available methods, it is suggested that those in which geomorphological surveys provide a first stage for environmental assessment may be particularly valuable.

Economic Geomorphology

Economic geomorphology and its significance can be linked to a growing recognition of the practical application of geomorphic principles and the findings of geomorphological research to human beings and the society at large who are influenced by and, in turn, influence the surface features of the earth. Rise in population has led to pressure on land resources, extension of agriculture to hilly and marginal lands resulted in anthropogenic catastrophes like soil erosion, landslides, sedimentation and floods. A proper interpretation of landforms throws light upon the geologic history, structure, and lithology of a region. As geology becomes more specialized there is growing possibility that the application of geomorphology to problems of applied geology will be overlooked. Human beings have over time tried to tame and modify geomorphic/environmental processes to suit their economic needs and at times the natural geomorphology of a place favors economic situation. Geomorphology has diverse application over a large area of human activity while geomorphologist may serve more effectively the need of society so both geomorphology and geomorphologist have economic importance.This can be understood better if one looks at the diverse role of geomorphology and its linkages with other fields.

Geomorphology And Hydrology

Water either on the surface of the earth or groundwater used by human is available from different sources like streams, lakes and rivers. The human civilizations chose sites based on availability of water. For domestic, commercial or industrial use, the presence of water plays a significant role in the economic development of region. The presence of water is dependent on the different lithological zones with varying conditions of surface as well as groundwater.

Hydrology of Limestone Terrains

Profound understanding of geomorphology is imperative to comprehend the hydrological importanceof the limestone terrain. Availability of water in limestone region depends on the type of rock. On the basis of permeability, limestone rocks may be primary or secondary. Calcareous sediments decide the formation of rock and its primary permeability while earth movements in the form of tension and compression such as faulting, folding, warping, and due to solution or corrosion mechanism decide the secondary permeability.

Geomorphology plays an important role to obtain water in limestone region. It may be easy or difficult to obtain water from wells in a limestone terrain. If the limestones have enough permeability and are capped with sandstone layer, the yield of water may below or inadequate, but subject to contamination. Karst plains lacks filtering cover, and any swallow holes, sinkholes, or karst valleys within an area of clastic rocks should cast doubt upon the purity of the water of springs.

Glaciated Areas and Groundwater

Preglacial and glacial time history; types of deposits and landforms determine the possibilities of large supplies of groundwater potentials in glaciated regions. There may be yield of large volume of water from Outwash plains, valley trains, and intertill gravels or buried outwash. Due to clay content most of the aquifers are poor, but containing local strata of sand and gravel may hold and supply enough water for domestic needs. The study of preglacial topography and geomorphic history of the area can help detect the presence and absence of underground water.

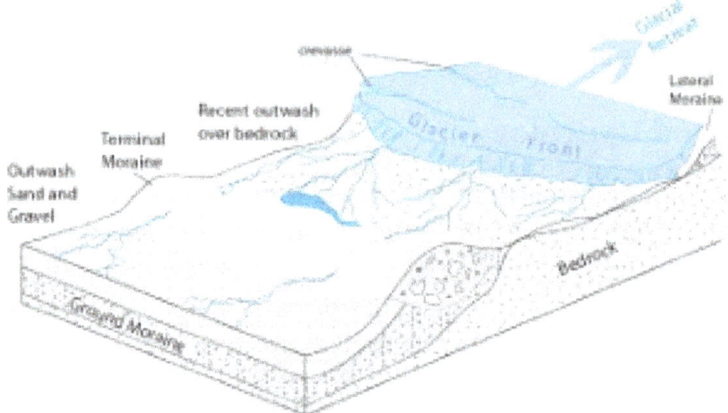

Geomorphology and Mineral Exploration

Minerals have long been an important resource for economic development of any region. There is a close association of geomorphology, geological structure and minerals deposits. Characteristic of landscapes of specific areas could indicate these geological structures. Economic geologist has not appreciated the exploration of some minerals in the name of understanding of the geomorphic features and history of a region. In search for mineral deposits, these three points may serve for Geomorphic features as:

(1) Some mineral have direct topographic expression for its deposits;

(2) The geologic structure and topography of an area have correlation which clue the accumulation of minerals;

(3) Geomorphic history clearly indicates the physical condition under which the minerals accumulated or were enriched in a particular area.

Surface Expression of Ore Bodies

Some of ore bodies have surface expression, but many do as topographic forms, as outcrops of ore, gossan, or residual minerals, or as such structural features as faults, fractures, and breccia zones. It is not necessary that all ore outcrops are reflected in positive topographic forms. The lead-zinc lode could be marked by a conspicuous ridge in the caseof Broken Hill, Australia. Quartz veins could stand out prominently as they are much more resistant to erosion than the unsilicified rocks, as in Chihuahua, Mexico. Some veins and mineralized areas may lack conspicuous topographic expression or it may be reflected by subsidence features or depressions. Though no generalization can be made about the exact type of topography necessary for the iron ore accumulation, distinct topographic expression is needed for a particular deposit. Residual iron deposits are the results of concentration of iron due to long periods of weathering, and thus for their accumulation, old erosion and weathering surfaces are favorable sites.

Weathering Residues

Geomorphology can play an important role for several important economic minerals, which are essentially weathering residues of present or ancient geomorphic cycles. Apart fromiron deposits, materials like clay minerals, caliche, bauxite and some manganese and nickel ores are of this nature. Recent weathering surfaces may exhibit residual weathering products or it may lie upon ancient weathering surfaces which are now buried.

Peneplain or near peneplain surfaces are most commonly surfaces upon which they form. In general,such minerals are to be found upon remnants of tertiary erosional surfaces above present base levels of erosion. It is not yet clear why the weathering of igneous rocks produces both clay minerals or hydrous aluminum silicates and hydrous oxides of aluminum, such as bauxite. It could be the varying climatic condition under which weathering occurs, that determines the final compound. The residual products from the weathering of igneous rocks are clay minerals found in temperate climates known as kaolinization. It should be recognized that numerous minerals other than kaolin may form in same climate. On the contrary, under tropical climates final weathering products are hydrous oxides of such metals as aluminum, manganese and iron. This type of weathering is known as laterization. The phase of geology which concerns with the recognition and the study of ancient weathering surfaces and soil has come to be known as paleopedology. Though it offers many possibilities but it is still in its infancy in the search for the type of mineral deposits designated as weathering residues of geological phase.

Placer Deposits

Placer deposits are mixtures of heavy metals within specific location and distinctive topographic expression. Geomorphic processes are the main cause of placer concentration of minerals.The deposition of placers is affected by the type of rock forming the bedrock floor. There are as many as nine types of placer deposits. They are residual or 'seam diggings', alluvium, colluvial, eolian, bajada, beach, glacial including those in end moraines and valley trains, and buried and ancient placers. The most important among them is alluvial placers.

Residual placers are also termed as 'seam diggings' which are residues from the weathering of quartz stringers or veins, are usually of partial amount, and grade down into lodes. Creep down slope is the main reason for the production of colluvial placers and these are transitional between residual placers and alluvial placers.

Colluvial placers (the koelits) and alluvial placers (the kaksas) are parts of the tin placers of Malaya. About one-third of the world's platinumin Russia, Colombia, and elsewhere, has been obtained from alluvial placers. The most important minerals like gold, tin and diamonds are obtained from alluvial placers. Most of the gold placers of this form have beenfound in California, Australia, New Zealand, and elsewhere. South Africa's diamonds from Vaal and Orange River districts, the Lichtenburg area, the Belgian Congo, and Brazil's Minas Geraes, are obtained from alluvial placers. Placer deposits have total share of around 20 per cent of world's diamonds. Australia, lower California, and Mexico have yielded gold in aeolian placers. Gold in California and Alaska, diamonds in the Namaqualand district of South Africa, zircon in India, Brazil, and Australia, and ilmenite and monazite from Travancore, India have yielded from beach placers.

Oil Exploration

Several oil fields have been discovered because of their striking topographic expression. These oil fields are characterized by anticlinal structures which standout distinctively in the landscape.In aerial photographs, many of the Gulf Coast salt dome structures are evident in the topography. For the student of geomorphology, it is fairly good working principle to suspect that areas that are topographically high may also be structurally high, where possibilities of topographical inversion at the crest of a structural high may result with weak beds.

In regions of dense tropical forest, topography cannot be seen through theforest cover, an anticlinal or domal structure may outline due to the tonal differences in the vegetation. In search for oil, more subtle evidence of geologic structures favourable to oil accumulation is being made. Aerial photographs are being used for analyzing drainage in typical terrain. Drainage analysis is useful particularly in regions where rocks have low dips and the topographic relief is slight. Permeability may be either primary or secondary in carbonate rocks. Oil yields from limestone may be in high quantity from rocks, which have a high degree of permeability produced by solution.

Elongate buried sand bodies are basically shoestring sands. Probably there is no phase of petroleum exploration which canbeusedto better advantage a knowledge of the in-depth characteristics of specific topographic features than that which deals with the misuse of shoestring sands. Most of the oil and gas sources are associated with unconformities -ancient erosion surfaces; hence a petroleum geologist must deal with buried landscapes.

Geomorphology And Engineering Works

Evaluation of geologic factors of one type or another are involvedin most of the engineering projects, amongst which terrain characteristics is most common. A detailed study of the geomorphic history of an area may support the proper evaluation of surficial materials and the bedrock profile configuration.

Road Construction

Topographic features of an area determine the most feasible highway route. Road engineering faces a number of problems due todifferent types of terrain that includes geologic structure, geomorphic history of the area, lithological and stratigraphic characteristics and strength of the surficial deposits. Areas like karst plain require repeated cut and fill and if not properly carried out, the road will be flooded after heavy rains with surface runoff from the sinkholes.

The presence of enlarged solutional cavities in karst region makes it necessary for the road to be designed in such a way thatit is not weakened. Regionslike glacial terrain presents a number of engineering problems. Road construction in flat till plain is topographically ideal but other areas where moraines, eksers, kames or drumlins like features exist there is need for cut and fill to avoid circuitous routes. Areas which are characterized by late, youth and maturity of relief will require more bridge construction and many cuts and fills. These types of areas are consistently facing problems like landslides, earth flows, and slumping.

Landslides and different types of mass wasting present problems not only in different phases of engineering but in highway construction also. Subgrade or the soil beneath a road surface has become more significant because of its control over the drainage beneath a highway, therefore construction design of highway should be in such a way to carry heavy traffic. Two factors largely determine the lifetime of a highway under moderate loads is the quality of the aggregate used in the highway and the soil texture and subgrade drainage. The type of parent material and the relationships of soils to its varying topographic conditions are more essential in modern road construction.

The most serious problems encountered by highway engineers is Pumping which means expulsion of water from beneath road slabs through joints and cracks. It is evident that pumping is particularly greater over glacial till than over permeable materials such as wind-blown sand and outwash gravel. Poor drainage in a subgrade is mainly responsible for pumping. Poor and best performance of the highway is characterized by silty-clay subgrades with a high water table and granular materials with a low water table respectively.

Dam Site Selection

A synthesis of knowledge concerning the geomorphology, lithology, and geological structure of terrains has greatly helped while selecting sites for dam construction. According to Bryan, five main requirements of good reservoir sites depend on geologic conditions:

1) Adequate size water-tight basin;
2) A Narrow outlet of the basin with a foundation that will permit economical construction of a dam;
3) To Build an adequate and safe spillway to carry excess waters;
4) Availability of resources needed for dam construction (earthen dams); and
5) Assurance that excessive deposition of mud and silt will not shortenthe life of reservoir.

Constructing a dam in a limestone terrain may prove a difficult one, for instance, the Hondo reservoir was built over limestonein southeastern New Mexico, while the water table was about20feet below the surface. Rapid leakage was the cause of abandonment of the reservoir. Building a dam in a valley may not be a good site from the standpoint of the size of the dam. Buried bedrock valleys containing sand and gravel fills are common in glaciated areas, which may not depict adequate picture of surface condition. Making dam on those sites where subsurface topography is not supportive with buried preglacial valley with sand and gravel in it would have a chance of leakage.

Location of Sand and Gravel Pits

Sand and gravel have more commercial and industrial uses. Evaluation of geologic factors such as variation ingrade sizes, lithologic composition, degree of weathering, amount of overburden, and continuity of the deposits are important while selecting suitable sites for sand and gravel pits. Floodplain, river terrace, alluvial fan and cone, talus, wind-blown, residual, and glacial deposits of various types are areas where sand and gravel may be found in abundance. In recent years, there is a great demand of gravel than sand due to decreased use of plaster in home construction therefore knowledge of various grade sizes is more important.

There are high proportions of silt and sand in floodplain deposits which show many variable and vertical gradations and heterogeneous lateral. With their angular shape as well as variable in size alluvial fan and cone gravels are found near their apices. Being angular like talus materials are too large to be useful and are limited in extent. There is only sand in wind-blown sands but have no gravel. Residual deposits are likely to contain pebbles that are suitable for cement work. These residuals are also limited in extent. Favorable sites for pits are terraced valley trains and outwash plains, which are usually extensive and do not have a thick overburden. Due to its large amount of material, kame deposits show a poor degree of assortment because it is discarded on the ground of being too large or too fine.

Geomorphology and Military Geology

During wars the military used information that was more geologic than geomorphic in nature. The information regarding digging trenches, mining, countermining, and water supply or other material was not utilized. Topography became more important during World War II with the development of the blitzkrieg type of warfare, because effectiveness of a blitz depends to a large extent upon the trafficability of the terrain. In recent years terrain appreciation or terrain analysis have become more important in military.

For a terrain if geological maps fail somewhere, geographic principlescan be applied with advantage to interpreting the terrain from aerial photographs. Little training is required to recognize features like mountains, hills, lakes, rivers, woods, plains or some kinds of swamps. It is important to know the kind of hill, plain, river or lake, and so on, because by knowing this it is quite possible to reconstruct the geography of that region. Aerial photographs are useful for the preparation of terrain intelligence as they provide information on the geomorphologyof the area. Terrain has been an important factor in the Korean War and in the fighting in Vietnam region.

Geomorphology and Regional Planning

Geomorphologic information can be utilized at various levels of planning. Combination of topographic information, soils, hydrology, lithology, terrain characteristics and engineering included on terrain maps make it suitable for regional planning. Applied geomorphology has distinct place in regional planning. On broad scale it can be used todelineate areas for forest, mountain, plateau, recreational, rural and urban areas. A balanced growth of a country's economy requires a careful understanding of its natural resources and human resources. Rural or underdeveloped terrain fulfills a variety of recreational needs. There is a transformation from a terrain maps into land-use suitability maps to develop rural and urban areas. Detailed information on topography enlighten-regional planners who may then advise development projects best suited for separate region.

Geomorphology and Urbanisation

There is a separate branch known as urban geomorphology applied to urban development. According to R.U. Cooke, this branch of geomorphology is concerned with "the study of landforms and their related processes, materials and hazards, ways that are beneficial to planning, development and management of urbanized areas where urban growth is expected". Geomorphic features decide the stability, safety, basic needs and even its expansion. That means city or towns entirely depend on lithological and topographical features, hydrological conditions and geomorphic features. Urban geomorphologist begin even before urban development through field survey, terrain classification, identification and selection of alternative sites for settlements irrespective of plain or hilly areas. These urban geomorphologists would be concerned with impact of natural events on the urban community and that of urban development on the environment.

When geomorphological problems are not understood by the planners and engineers then it leads to destruction and damage to urban settlements in different environmental regions. Settling of foundation material in dry or glacial region, weathering process, damages of roads and buildings through floods in many parts of the world occur frequently.These problems arise due to misunderstanding of the geomorphological conditions. In developing countries attention has not been given to the geomorphological conditions before the development of existing urban centres. This leads to haphazard growth of city with squatter settlement and shanty towns with urban morphology.

Geomorphology and Coastal Zone Management

Coastal zones are not linear as boundary between land and water isviewed as dynamic region of interface.The major threat to the fragile coastal zone is its deteriorating coastal environment through shoreline erosion, pollution and extinction of species coastal zone management requires an integrated approach. The most widespread material is beach sand, found mainly in low latitudes. Beach sand and gravel is widely used for construction industry.

Geomorphologists have made some significant contribution towards an understanding of shoreline equilibrium in Eastern Australia where considerable development of sand mining for heavy minerals has been done. Some structural measures have been designed forcoast protection such assea-defence structures -seawalls, breakwaters, jetties and groynes. To protect the sea backshore zone from direct erosion cut, sea walls are designed since these walls are impermeable they increase the backwash and produce a destructivewave effect. Breakwaters can be built either normal or parallel to the coast. It is necessary to monitor and quantify wave conditions, tidal currents and sediment movement in the nearshore zone to evaluate how sea defenses and other human-made structures affect shoreline equilibrium.

In context of coastal zone management Hails emphasizes that applied geomorphology must be concerned with quantitative and not descriptive research in order to obtain relevant and accurate data on (i) natural erosion and deposition rate (ii) at what rates and amount the sediment transport from river catchments to the near shore zone; (iii) variations in sediment composition and offshore distribution; (iv) sand supply sources and shoreline equilibrium; (v) interchange rate of sand between beaches and dune systems; (vi) the effects of constructing sea defenses; (vii) offshore sediment dispersal and the dredging effects of seabed morphology, sediment transport and wave refraction; and (viii) analysis of landform including topography of the near-shore zone, form of the continental shelf and of relict coast lines, particularly in terms of rock outcrops. Above investigation provides relevant baseline data needed for systematic planning process and monitoring programme sand devising land use scheme.

Geomorphology and Hazard Management

Hazards can be natural or human-induced where tolerable level or unexpected nature exceeds. According to Chorley, geomorphic hazard may be defined as "any change, natural or man-made, that may affect the geomorphic stability of a landform to the adversity of living things". These hazards may arise from immediate and sudden movements like volcanic eruptions, earthquakes, landslides, avalanches, floods, etc. Faulting, folding, warping, uplifting, subsidence, or vegetation changes and hydrologic regime due to climatic change arise fromthe long term factors. Areas having past case histories of volcanism and seismic events help in making predictions of possible eruptions and earthquakes respectively. Regular monitoring of seismic waves, measurement of temperature of craters lake, hot springs, geysers and changes in the configuration of volcanoes whether dormant or extinct can reduce the hazard to some extent. A detailed knowledge of topography can predict the path of lava flow and its eruptions points in advance.

The behavior of a river system can be well understood by its geomorphic knowledge through its channel morphology, flow pattern, river metamorphosis and so on. It may help controlling excess water in river and control measures during flood season. Prior knowledge of erosion in the upper catchment area and carrying sediments to its proportion may help in understandingthe gradual rise in river bed, which may lead to levee breachand sudden floods. Earthquakes may be human induced or natural geomorphic hazards. Detailed study of seismic waves help in identifying and mapping the zones of high to low intensity to reduce the risk of human life.

With the introduction of aerial photographs and satellite imageries, preparation of specialized maps and interpreting them has become easier and more accurate. Aerial photographs are being used for evaluating landforms and land use for city developmental plans, construction projects, highway etc. Another tool i.e. Remote sensing is necessary for sustainable management of natural resources like soil, forest, crops, oceans, urban and town planning etc. At present Geographical

Information Systems (GIS) technology has been used along with Remote Sensing techniques in geomorphic features interpretation.

By understanding the core of geomorphology and its interlinkages with other fields and how this interlinkage influences human beings and further the society one gets to know the important role played by geomorphology in contributing to the economy.Whether itis urbanization, mineral exploitation, oil exploration,water availability, construction of dams, roads, railway tracks etc.,or disaster management the role played by geomorphology is immense and imperative.

References

- What-is-geomorphology: wisegeek.com, Retrieved 04 April 2018
- De Vriend H.J., 1991. Mathematical modelling and large-scale coastal behaviour, Part 1: Physical processes. Journal of Hydraulic REsearch 29, pp. 727-740
- Geomorphology-explained, geography-environment: brightknowledge.org, Retrieved 14 May 2018
- Legendre, P., S.F. Thrush, V.J. Cummings, P.K. Dayton, J. Grant, J.E. Hewitt, A.H. Hines, B.H. McArdle, R.D. Pridmore, D.C. Schneider, S.J. Turner, R.B. Whitlatch & M.R. Wilkinson, 1997. Spatial structure of bivalves in a sand flat: Scale and generating processes. Journal of Experimental Marine Biology and Ecology, 216, pp. 99-128
- Geomorphology-202005: uky.edu, Retrieved 24 July 2018

2

Geomorphological Processes

Geomorphological processes are responsible for the creation and change to the topographic features. Some of these are tectonism, volcanism, groundwater movement, surface water flow, air movement, glacial action, etc. Such processes can be categorized under different groups such as aeolian, biological, fluvial, glacial and igneous processes, among others. The topics elucidated in this chapter cover some of these important processes for an in-depth understanding of geomorphology.

Aeolian Processes

Aeolian processes, involving erosion, transportation, and deposition of sediment by the wind, occur in a variety of environments, including the coastal zone, cold and hot deserts, and agricultural fields. Common features of these environments are a sparse or nonexistent vegetation cover, a supply of fi ne sediment (clay, silt, and sand), and strong winds. Aeolian processes are responsible for the emission and/or mobilization of dust and the formation of areas of sand dunes. They largely depend on other geologic agents, such as rivers and waves, to supply sediment for transport.

Areas of sand dunes occur in inland and coastal settings, where they often provide a distinctive environment that provides habitats for endemic and rare or threatened species. In both coastal and inland settings, dune migration and sand encroachment may impact neighboring ecosystems and resources, as well as infrastructure.

Transport of fine sediment by wind may cause dust storms, events in which visibility is reduced to less than 1 km by blowing dust. Dust storms impact air quality in their immediate vicinity as well as in areas downwind. Deposition of dust may have a significant effect on the composition and nature of soils in arid regions and beyond. Far-traveled dust from distant sources may have a significant effect on soil chemistry and nutrient status.

Transport of Particles by Wind

Movement of particles by the wind takes place by a combination of direct wind shear stress and atmospheric turbulence. There are three modes of sediment transport by wind: creep or reptation; saltation, and suspension (figure below). The mode of transport depends primarily on the ratio between particles settling velocity, and hence particle size, and wind shear stress and turbulence intensity. Very small particles (<20 microns) are transported in suspension (tens of km or greater) and are kept aloft by turbulent eddies in the wind. True suspension occurs when the particle settling velocity is very small compared to the turbulence intensity of the wind. Larger particles (20–70 microns) undergo short-term suspension for distances of tens to hundreds of meters; material of sand size (70–1000 microns) is transported mainly in a series of short hops (saltation), in which the vertical component of wind velocity (turbulence) has a minimal effect on particle trajectories. Material coarser than 500 microns in diameter (coarse sand) is transported on surface by reptation and creep. The modes of transport are interdependent: saltating sand particles eject silt- and clay-sized particles into the wind and impact coarse grains that are rolled along the bed.

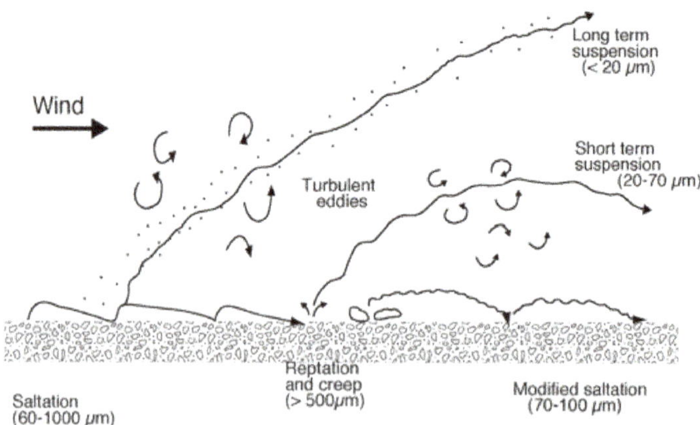

Figure. Modes of sediment transport by the wind

Grains begin to move and sediment is entrained by the wind when fluid forces (lift, drag, moment) exceed the effects of the weight of the particle, and any cohesion between adjacent particles as a result of moisture, salts, or soil crusts. The threshold wind speed at which grains begin to move is strongly dependent on particle size (figure below part A). For quartz sand, the minimum threshold velocity is associated with fi ne sand (~100 microns diameter). The mass flux or transport rate of sand has been determined by numerous laboratory wind tunnel and field studies to be proportional to the cube of wind shear velocity above a threshold value (figure below part 2). For any wind shear velocity, there is a potential rate of sand transport or transport capacity, which is only reached when the availability of sediment is unrestricted (e.g., most loose sand surfaces). In these conditions, the wind is saturated with respect to transport capacity.

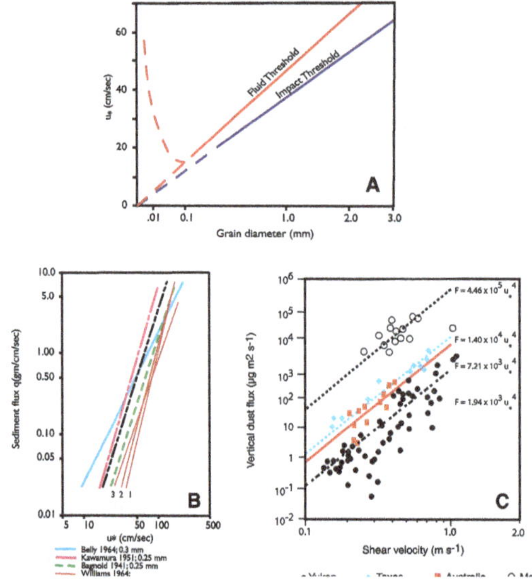

Transport of sediment by the wind: (A) Relation between threshold wind shear velocity and particle size. (B) Mass flux of sand as a function of wind shear velocity. Data from laboratory wind tunnel experiments. (C) Relations between horizontal flux of sand-sized particles and vertical flux of dust. Data from field experiments.

Very fine grains (silt and clay size) are inherently resistant to entrainment, yet are readily transported by the wind. Recent studies have shown the critical role of impacting sand grains in the mobilization of silt- and clay-size particles and demonstrated the close relations between the horizontal flux of sand-size particles and the vertical flux of fi ne particles. In these situations, the horizontal mass transport rate is directly related to shear velocity (figure 2B), so dust emissions scale to the fourth power of wind shear velocity (figure above part C). Where there is a limited supply of particles able to abrade soil clods or playa crusts, dust emissions are limited by the supply of particles rather than the wind shear velocity, and the vertical flux of dust is almost independent of wind shear velocity.

Wind Erosion

Wind Erosion by wind involves two linked processes: abrasion (mechanical wearing of coherent materials, including playa crusts and clods created by tillage) and deflation (removal of loose material). Considerable attention has been devoted to the processes and rates of wind erosion because of their impact on agriculture, especially in semi-arid regions, and the implications of dust emissions for air quality. Wind erosion abrades crops, removes organic matter, nutrients and fertilizer, and changes soil texture. The products of wind erosion (especially dust particles) impact air quality, atmospheric radiative properties, and human health, causing respiratory illnesses. Rates of wind erosion vary widely and for a given wind shear velocity are dependent on soil or sediment texture and the degree of crusting and cohesion. The highest emission rates for fi ne-grained sediment are associated with soils of loamy texture, especially those that have been disturbed by vehicular traffic and/or animals.

Aeolian Deposits

Aeolian deposits include sand seas and dune fields, deposits of silt (loess), and fi ne-grained material that forms a significant component of desert margin and other soils.

Aeolian deposits—silt and clay size. Deposits of wind transported, silt-sized quartz particles, termed loess, cover as much as 10% of Earth's land surface. Loess deposits are widespread in areas of northern China, southern central Asia, central Europe, Argentina, Alaska, and the central United States. Much of the material was thought to be derived from silt particles produced by glacial grinding and supplied to aeolian processes by glacial outwash ("glacial loess"), but other processes, including frost shattering, salt weathering, reduction in size during transport by rivers, and aeolian abrasion are important, especially in the formation of "desert loess."

Silt- and clay-sized material of aeolian origin is also an important component of many desert margin soils. Deposition of silt plays a role in the formation of many stone pavement surfaces in desert regions (desert pavement). These surfaces are characterized by a surface layer of gravel or larger clasts (particles) that overlie fi ne-grained materials. Detailed studies of these surfaces show that the surface layer of gravel rests on a layer of soil-modified dust that may be a meter or more thick and mantles bedrock or materials deposited by other processes (e.g., alluvial sediments). The dust is trapped by the clasts and deposited between them. The fi ne material is incorporated into the mantle by the shrinking and swelling of clay minerals so that the clasts remain at the surface as they inflate over periods of thousands of years.

Aeolian deposits—sand dunes. Aeolian dunes form part of a hierarchical system of self-organizing bed forms which comprises: (1) wind ripples (spacing 0.1–1 m); (2) individual simple dunes or superimposed dunes on compound and complex dunes (spacing 50–500 m); and (3) compound and complex (mega-) dunes or draa (spacing more than 500 m). Dunes occur wherever there is a sufficient supply of sand-sized sediment, winds to transport that sediment, and conditions that promote deposition of the transported sediment. These requirements are satisfied in two main environments: (1) coastal areas with sandy beaches and onshore winds; and (2) desert areas. Most dunes occur in contiguous areas of aeolian deposits called sand seas (>100 km2) or dune fields.

Wind ripples (figure) typically have a wavelength of 0.05– 0.2 m and amplitude of 0.005–0.010 m. They are ubiquitous on sand surfaces, except those undergoing very rapid erosion or deposition, and form because a flat sand surface over which sand transport by saltation and reptation occurs is dynamically unstable.

Figure: Wind ripples, Wind direction from left to right.

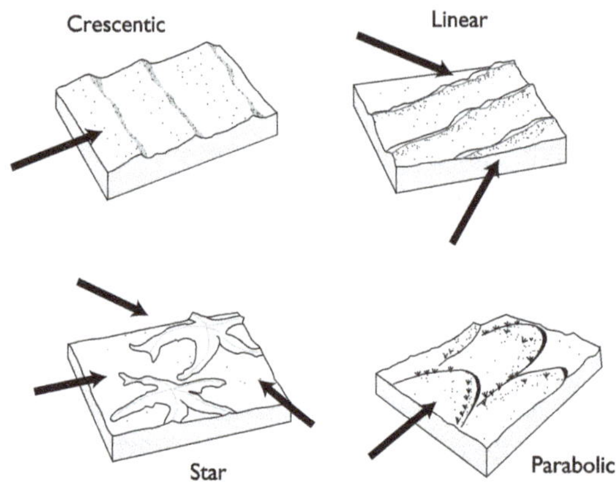

Figure: Major dune types

Aeolian dunes occur in a self-organized pattern that depends on the wind regime (especially its directional variability) and the supply of sand. Sand dunes occur in four main morphologic types (figure above): Crescentic (transverse), linear, star, and parabolic. The simplest dunes form in areas characterized by a narrow range of wind directions. In the absence of vegetation, crescentic dunes will be the dominant form. Isolated crescentic dunes or barchans occur in areas of limited sand supply, and coalesce laterally to form crescentic or barchanoid ridges as sand supply increases. Linear dunes are characterized by their length (often more than 20 km) sinuous crest line, parallelism, and regular spacing. They form in areas of bimodal or wide unimodal wind regimes. Star dunes have a pyramidal shape, with three or four sinuous sharp-crested arms radiating from a central peak and multiple avalanche faces. Star dunes occur in multidirectional or complex wind regimes and are the largest dunes in many sand seas, reaching heights of more than 300 m. Parabolic dunes are characterized by a U or V shape with a "nose" of active sand and two partly vegetated arms that trail upwind. They are common in many coastal dune fields and semi-arid inland areas, and they often develop from localized blowouts in vegetated sand surfaces. Other important dune types include nebkhas, or hummock dunes, anchored by vegetation (common in many coastal dune fields); lunettes (often composed of sand-sized clay pellets) that form downwind of small playas; and a variety of topographically controlled dunes (climbing and falling dunes, echo dunes).

Relations between dune types and wind regimes indicate that the main control of dune type is the direction of the wind. Grain size, vegetation cover, and sediment supply play subordinate roles in desert areas. In semi-arid and coastal areas, vegetation cover plays a major role in Aeolian dynamics.

Stressors and Possible Changes

The state of an aeolian geomorphic system is controlled by the supply of sediment of a size suitable for transport by the wind; the mobility of the supplied sediment, which is controlled by wind conditions; and the availability of sediment for transport, determined by vegetation cover and soil moisture. Changes in these external drivers can be the result of climate or human impacts. Climate change and variability affects the mobility of sediment through variations in wind strength; vegetation cover and soil moisture are directly influenced by the amount of precipitation; the supply of sediment may be affected by changes in wave energy, beach sediment budgets, or river discharge. Changes to aeolian systems that can be attributed to the effects of climate variability on annual to decadal time scales include changes in the magnitude and frequency of dust storms, sand transport rates, and activation or stabilization of areas of sand dunes. Such changes are a good indication of the response of a landscape to drought periods. In addition, human impacts may affect vegetation cover by grazing pressure or trampling by animals or people, and increase sediment availability of soils due to disturbance by animals or off-road vehicles. Humans can also directly or indirectly affect sediment supply from rivers or the coastal zone.

Biological Processes

Living organisms can interact in a significant way with landforms and such interactions play a determining role in the shaping and morphing of the Earth's topology. These processes can be termed as bio-geomorphologic processes. Such biological processes can have varied manifestations. Biogeochemical processes can control chemical weathering while mechanical processes such as burrow and tree throw formations influence soil development. Global erosion rates can be manipulated using climate modulation through carbon dioxide balance. Trees and grasses act as stabilizers by maintaining shorelines and limiting wind erosion. The chemical interactions existing between plants and rocks are responsible for the breaking and splitting of rocks organically over time. Animals also have an effect, often as a removal of vegetation, which further allows other interactions with the landscape. It is thus rare to find terrestrial landscapes on which biological influence is insignificant or non-existent.

Fluvial Processes

Fluvial process is the physical interaction of flowing water and the natural channels of rivers and streams.

Such processes play an essential and conspicuous role in the denudation of land surfaces and the transport of rock detritus from higher to lower levels.

Over much of the world the erosion of landscape, including the reduction of mountains and the building of plains, is brought about by the flow of water. As the rain falls and collects in watercourses, the process of erosion not only degrades the land, but the products of erosion themselves become the tools with which the rivers carve the valleys in which they flow. Sediment materials eroded from one location are transported and deposited in another, only to be eroded and redeposited time and again before reaching the ocean. At successive locations, the riverine plain and the river channel itself are products of the interaction of a water channel's flow with the sediments brought down from the drainage basin above.

The velocity of a river's flow depends mainly upon the slope and the roughness of its channel. A steeper slope causes higher flow velocity, but a rougher channel decreases it. The slope of a river corresponds approximately to the fall of the country it traverses. Near the source, frequently in hilly regions, the slope is usually steep, but it gradually flattens out, with occasional irregularities, until, in traversing plains along the latter part of the river's course, it usually becomes quite mild. Accordingly, large streams usually begin as torrents with highly turbulent flow and end as gently flowing rivers.

In flood time, rivers bring down large quantities of sediment, derived mainly from the disintegration of the surface layers of the hills and valley slopes by rain and from the erosion of the riverbed by flowing water. Glaciers, frost, and wind also contribute to the disintegration of the Earth's surface and to the supply of sediment to rivers. The power of a river current to transport materials depends to a large extent on its velocity, so that torrents with a rapid fall near the sources of rivers can carry down rocks, boulders, and large stones. These are gradually ground by attrition in their onward course into shingle, gravel, sand, and silt and are carried forward by the main river toward the sea or partially strewn over flat plains during floods. The size of the materials deposited in the bed of the river becomes smaller as the reduction of velocity diminishes the transporting power of the current.

Since the earliest days of modern applied hydraulics, engineering research has attempted to better understand sediment transportation. Because sediment particles are generally heavier than the amount of water they displace, the Archimedes principle could not be used to explain the fact that heavy sediment was capable of being lifted and transported by flowing water. Twentieth-century research distinguishes, in this connection, between "bed load" on the one hand and "suspended load" on the other. The former is composed of the larger particles, which are either rolled or pushed along the bed of the stream or which "jump," or saltate, from the crest of one ripple to another if the velocity is sufficiently great. On the other hand, the smaller particles, the suspended sediment once picked up and lifted by the moving water, may remain in suspension for considerable periods of time and thus be transported over many kilometers.

The Main Processes of Fluvial Erosion

- Abrasion: The erosion of the river bottom and the riverbank by material being carried by the river itself.
- Attrition: The rocks and pebbles being carried by the river crash against each other, wearing them down to become smaller, rounded pebbles.
- Corrosion: The chemical erosion of the rocks of the riverbank by the slightly acidic water. This occurs in streams running through rocks such as chalk and limestone.
- Hydraulic Action: The water forces air to be trapped and pressured into cracks in the rocks on the bank of the river. This constant pressure eventually causes the rocks to crack and break apart.

Fluvial Transportation

Once it has been eroded, material in the river is transported down the river. Whilst this is happening, erosion processes such as attrition and abrasion continue to occur. There are four main processes of fluvial transportation, depending on the size of the material being moved.

- Traction: The largest rocks in the river are slowly rolled along the bottom of the river by the force of the water. This primarily occurs in the upper reaches of the river.
- Saltation: Smaller rocks are bounced along the riverbed. This occurs in the upper and middle sections of the river in general.
- Suspension: The water carries smaller particles of material. This process occurs throughout the course of the river, but increases the closes you are to the mouth of the river.
- Solution: Material is dissolved within the water and carried along by it. Primarily this occurs in the middle and lower reaches of the river.

Fluvial Deposition

Fluvial deposition occurs where the river losses energy and therefore cannot continue to carry the material it is transporting. This could happen in an estuary when the river meets the sea and slows down, depositing its load, which may eventually lead to the formation of salt marshes or a delta. Material is also deposited further up the course of the river. For instance the slower moving water on the inside of a bend of a river will have less energy and therefore drop its load, helping to create a meander.

A major depositional feature of a river is the flood plain, in its lower reaches. This is made up of deposited sand and silt, which is known as alluvium. This is often very fertile and is the reason why many areas near rivers have large amounts of agricultural activity.

Glacial Processes

Glacial processes involving erosion, transportation, and deposition of sediment by the glaciers.

Glacial erosion of bedrock surfaces, intact bedrock units and sediments involve a range of processes, at times referred to as wear and attrition, that require a broad grasp of several closely allied components such as bedrock, glacial ice, glacial meltwater, sediment conditions and pre-glacial bedrock conditions. Bedrock erosion can occur depending on lithology, bulk shear strength, fracture hardness, penetration toughness, moisture content, bedrock structure, fracture and joint geometry, and temperature.

Many parameters such as penetration toughness and fracture hardness are largely unknown for most bedrock types. The part played by the glacier ice in erosive processes is fundamental in terms of ice thickness, basal stress levels, basal ice velocity, the basal ice debris content and the temperature at the erosive interface.

In terms of meltwater velocity and discharge events, the sediment content of the water, its temperature, glacial meltwater is a very effective erosive agent. Glaciated terrains exhibit superb examples of meltwater erosion in terms of P-forms, water abraded surfaces, tunnel valleys and networks. The depth of meltwater penetration below and active ice mass is a fundamental question when considering for example the depth of burial of nuclear waste. To date, evidence for hydromechanical changes caused by glaciation are limited and inconclusive. Estimates of the penetrative effect of glaciation range from several 10 s of metres to depths ≥ 300 m.

Until relatively recently, the role played by sediment as an erosive agent was largely ignored. The effectiveness of many erosive process whether by meltwater or direct ice contact is dependent to a great degree on the sediment acting at the erosive interface in terms of the sediment's lithology, and hardness. For already deposited sediment, it becomes essential to understand the geotechnical nature of each lithofacies type (field data that are presently generally limited beyond engineering results, are woefully inadequate. Such data take little or no account of structural geometry or clast/boulder content when measuring bulk geotechnical components) to understand how sediment can be glacially eroded.

From anecdotal evidence, it is suggested the rates of glacial erosion vary enormously from one glaciated basin to the next. Erosion rates can be characterized through the detrital geochemistry of sediments found in deltas or submarine tunnel valley infills beyond ice margins. In terms of ice sheet glacial erosion rates there is a dearth of data. The general average rates of erosion suggested are in the range of 0.07–30 mm a^{-1} for valley glaciers. In areas of rapid rock uplift, a value of 1.0–100 mm a_{-1} for valley glaciers. In areas of rapid rock uplift, a value of 1.0–100 mm a^{-1} 1 has been suggested suggested that rates under the Cordilleran Ice Sheet in western Canada were in the range of 0.09–0.35 mm a^{-1} but since beneath an ice sheet overlying rugged topography it is unlikely to represent the kind of values beneath other Pleistocene ice sheets. From East Antarctica erosion rates of 0.001–0.002 mm a^{-1} have been reported whereas, in comparison, rates across West Antarctica although sparse and varied are often much higher. Rates in the Pine Island Glacier region are 0.6 0.3 m a^{-1}, with an average current erosion rate of ~ 1ma^{-1}. This may be a function of topography and bedrock type although again the latter varies enormously across the ice sheet. Evidence from Antarctic a would suggest that much of the Antarctic interior may have been subject a total loss through erosion of <200 m of erosion.

The depth and degree of erosion reflects the presence or absence of warm-based ice and the consistency of ice flow direction. Near the continental margins selective linear erosion has over deepened pre-existing relief. Terrains close to ice streams have largely survived unmodified by glacial erosion if under cold-based ice. High erosion rates appear to result from steep thermal gradients in basal ice where basal ice velocity is high and warm-based ice prevails. It has been reported that erosion rates vary from 0.001 mm a^{-1} for polar cold-based ice and thin temperature plateau glaciers on crystalline bedrock to 0.1 mm a^{-1} for temperate, warm-based ice masses on resistant crystalline bedrock surfaces in Norway and Switzerland. Whereas in Alaska under fast warm-based ice masses rates of 10–100 mm a^{-1} have been noted. These considerable variations reflect bedrock lithology, ice basal thermal states and the basal ice velocity, and pre-glacial bedrock and topographic settings. For both the Laurentide Ice Sheet (LIS) and the Fennoscandian Ice Sheet (FIS) rates of estimated erosion vary enormously. Estimates for both the LIS and FIS tend to use a global glacial cycle as a measure of the amount of erosion with rates ranging from 0.2 to 4–5 m but typically in the range of ~ 1 mm a^{-1} . Based upon physical evidence and theoretical considerations it is only too apparent that, in general, where ice sheets are cold-based minimal erosion occurs, in contrast to where warm-based ice, moving fast especially in ice stream pathways then erosion rates are appreciably higher.

Glacial Transport—Processes

Transport pathways in ice sheets and glaciers are well established. However, englacial and supraglacial transport identification in Quaternary and Pre-Quaternary sediments and rocks remains precarious at best. In subglacial environments, the development and activity of sediment fluxes transporting sediment toward the ice margin needs to be fully explored and understood. In models of subglacial deforming sediment, the sediment flux of subglacial sediment is mobilized as a function of strain, pore water pressure and effective stress levels, and is seen as a consistent supply source. Subglacial sediment supply is a function of the production of "fresh" and "scavenged" sediment often sourced from areas remarkably close to the site of deposition (<15km). For example, beneath the paleo-ice stream draining Marguerite Bay in West Antarctica, Dowdeswell et al. calculated a sediment flux rate, assuming a 1 m thick mobile subglacial sediment layer, of 4–28 km^3 /1000 a^{-1} along a 35 km wide paleo-ice stream flowing at an ice velocity of approximately 0.5^{-1} km a^{-1}. These sediment flux rates equate to approximately 100–800 m^3 a^{-1}.

Glacial Deposition—Processes

The mechanics of glacial deposition has been well studied for over a century, but issues persist in terms of the specific processes involved with many aspects of glacial deposition and allied bedform/landform development. Many general hypotheses on deposition are well established now only refinement is required, yet there are still further revelations in terms, for example, of the process of immobilization of subglacial sediments into till sheets or plains or a specific range of bedforms loosely classified as MSGL.

Hillslope Processes

Hillslopes essentially cover the whole landscape. Some are steep, others are gentle. Under the natural condition, water, sediments, and rocks move down slope. The present topography and soil thickness reflect a sensitive balance between weathering, erosion, and deposition. Human activities can upset the balance and cause;

- Loss of the fertile soil from cultivated lands.
- Excessive erosion of lands and sediment load in streams.
- Catastrophic disaster of landslides.

We need to understand geomorphic processes caused by the action of water on hillslope.

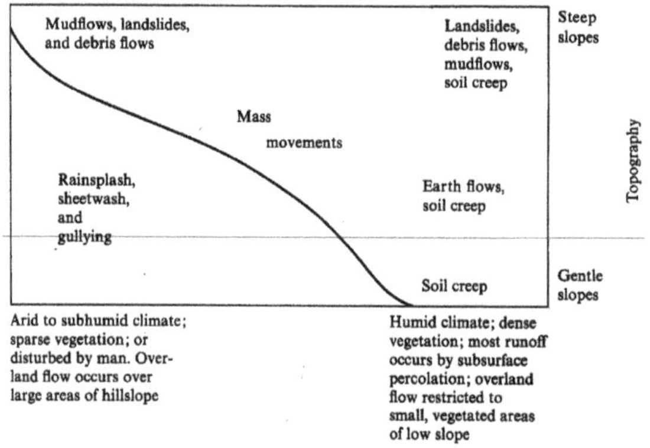

Conditions of climate, vegetation, land use, and topography under which various geomorphic processes dominate hillslope erosion

Normal and Accelerated Erosion

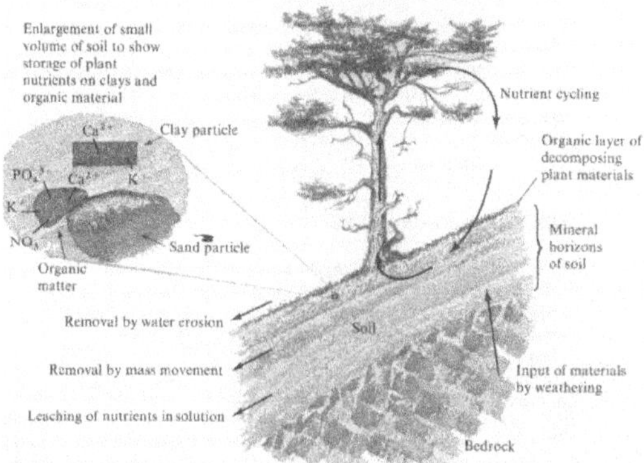

The soil achieves a balance between the input and output of materials

Soil profile develops over a long time. It reflects the balance of input (weathering of bed rock) and output (erosion). Normal rate of erosion keeps the soil profile stable under the current climate. Human activities can significantly accelerate erosion resulting in the soil loss.

Hillslope Erosion by Water

Erosion is driven by Horton overland flow, which primarily occurs on non-vegetated surface. Saturation overland flow does not contribute much because the soil surface is protected by a vegetative cover. Horton overland flow causes; rainsplash and sheetwash erosion, and gullying.

Rainsplash Erosion

A moving object posses kinetic energy (e.g. a car on a highway). This energy is released when the object hits something (highway clash). Each rain drop has kinetic energy which is released to splash soil particles upon hitting the ground. On average, the travel distance of a particle landing on the down-slope side is greater than the opposite, resulting in net downward motion of the soil. Each drop splashes a little bit, but billions of drops can cause significant erosion.

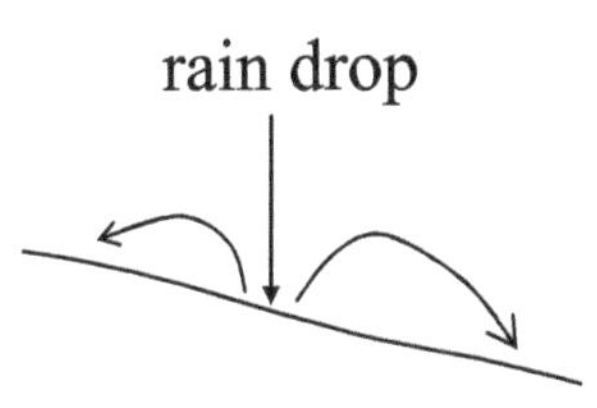

High-speed photograph of a raindrop impact on a soil surface

This process is particularly important on steep slopes devoid of vegetation. Thick vegetation protects soils from direct hits of rain drops. Soils are resistant to erosion when they have high organic content and reasonable amount of clay to develop aggregated structure. Tillage breaks down the structure and reduce the resistance to erosion and also the infiltration capacity.

Sheetwash Erosion

Horton overland flow results in an irregular sheet of water flowing downhill. The depth and velocity of flow increase downslope as more water is generated by precipitation excess. If the soil resistance to erosion is constant, the intensity of erosion depends on the product of water depth and hillslope gradient. In general, longer and steeper hillslopes has more intense erosion. Note that a sheet of water actually consists of many tiny streams and threads of water. We call it a sheet because, on long-term average, the erosion removes a roughly uniform depth of soil. Sheetwash erosion occurs with rainsplash erosion, and it is difficult to separate the two. Both are governed by the same factors; vegetation, slope gradient, and resistance of soil to erosion.

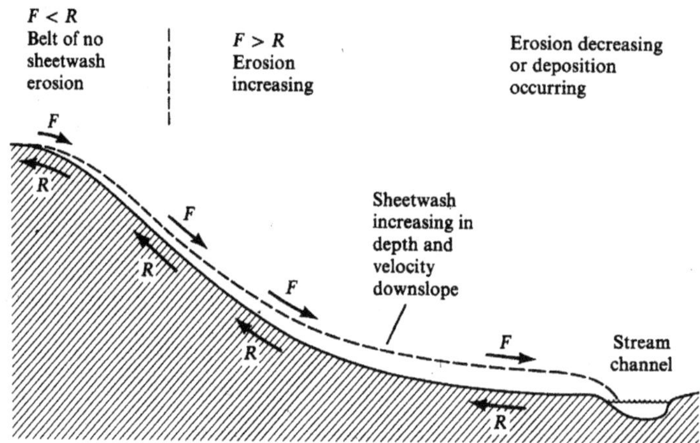

The relative magnitude of the eroding force of sheetwash (F) and the resistance (R) of the soil to erosion.

Rill and Gully Erosion

If the minute streams of water cut separate channels, the process is known as rill erosion. Water is concentrated in rills, and efficiency and intensity of erosion increases. When the depth of a rill exceeds about 0.3 m, it is called gully. Gully erosion produces incisions up to 101 m deep and 103 m long. Gullies occupy relatively small area. Contribution of gully erosion to overall erosion from a catchment is small (a few percent), but it has major localized effects.

Rates of Water Erosion

Thousands of experimental plots across Canada and US are used to determine the relationships between the erosion rate and various factors affecting erosion. These plots are equipped with water and sediment traps. Observations are also made in natural catchments using erosion pins.

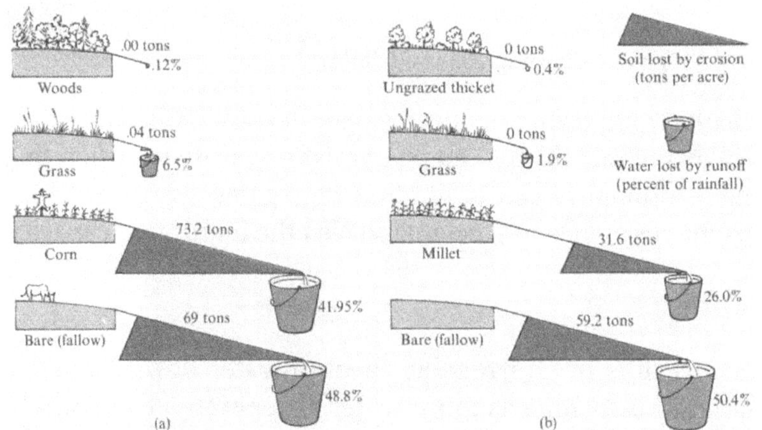

Runoff and erosion under various landuse. (a) Midwestern US. (b) Tanzania

Prediction of Soil Erosion

Planners frequently need to estimate the rate of soil erosion. The best method is to have some local field data, but that's not always available. The Universal Soil-Loss Equation is commonly used to get a rough estimate.

Significance of Soil Erosion

Soil erosion was recognized as a major threat to the continued productivity of the land in 1930's. Widespread use of fertilizers in US and Canada seemed to solve this problem, but it created other problems such as lake eutrophication. Also, some countries cannot afford applying enormous amount of fertilizers. Overall, the soil erosion must be avoided to achieve sustainable development.

Population growth in tropical and subtropical countries generates pressure on food production and energy sources (i.e. firewood and charcoal), which accelerates deforestation. Once the soil is lost, the land cannot sustain plant growth and desertification results.

Soil erosion increases sediment load into streams, which endangers aquatic species and fills up reservoirs used for controlling floods and generating power.

Control of Soil Erosion and Sedimentaion

(1) Agricultural lands

- Increase vegetative cover by adopting crop rotation that includes plants that provide a good cover and improve soil structure (e.g. alfalfa).
- Avoid tilling as much as possible. Cultivate along topographic contour (contour cultivation) to increase surface detention capacity and promote infiltration.
- Plant grains and cover crop (alfalfa) alternately as strips along topographic contour (strip cropping).
- Cover the ground surface with hay, straw, or manure (mulching) to increase resistance and infiltration capacity.
- Divide the hillslope into small pieces of flat lands to reduce the length and slope gradient of each piece (terracing).

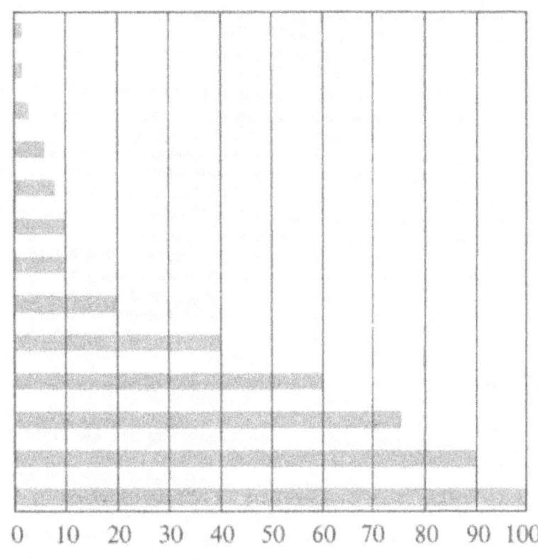

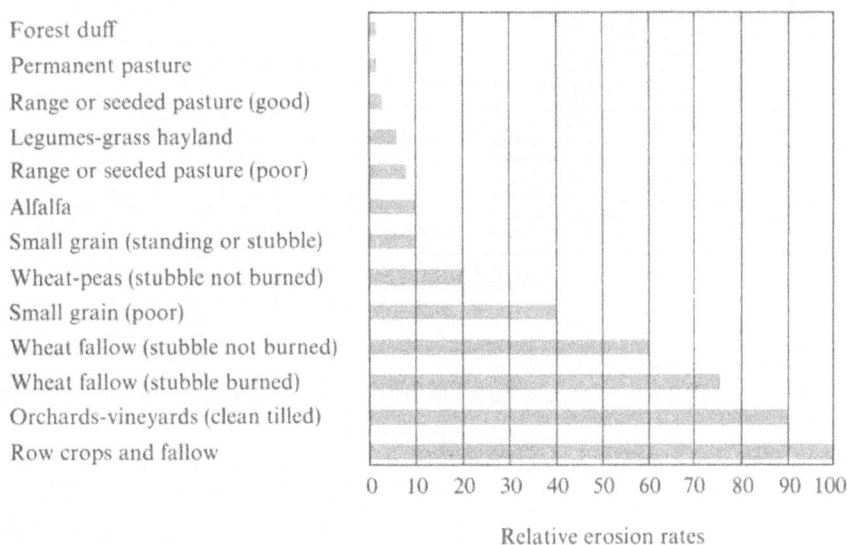

Relative erosion rates under different vegetal covers

(2) Rangelands

- Control the number of stock to meet with carrying capacity of the land.

(3) Forest

- Log the trees so that forest duff layer is not destroyed.
- Logging roads are the primary source of sediments. Try to layout logging roads on the gentlest gradient possible, preferably along ridges. Restore the area once logging operation is complete.

(4) Mine sites

- Restore the vegetation as much as possible, although this may be difficult in areas affected by acid mine drainage.

(5) Urbanland

- In construction sites, minimize the time the site is left unprotected. Maintain as much original vegetation as possible. Cover the area with surface mulch.
- Planning is very important. Classify the lands according to soil depth, texture, and land gradient. Develop the areas based on the land classification.

Mass Wasting

This mode of hillslope processes move enormous amount of mass, up to millions of m^3, and often quite rapidly.

- Damage of houses and structures
- Economic damage (lost farms, power lines, filled up reservoirs, etc.)
- Loss of human life

The hillslope failures are classified as follows.

Falls	Rock falls
Slides	Planner failures (rock slides, debris slides)
	Rotational failures (slumps)
Flows	Debris avalanches. Debris flows. Earthflows.
	Mudflows. Solifluction (arctic region).
Soil creep	

The classification is by no means clear-cut and combination of several processes may occur.

Rockfalls occur on very steep slopes. Talus slopes and screes develop under such slopes. Cost of maintaining highways in rockfall-susceptible areas is expensive.

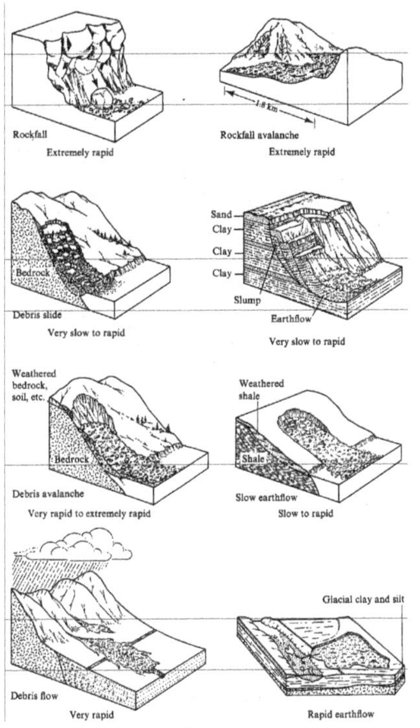

Types of hillslope failures

In slides the failure occurs along a surface or within a narrow zone of deformation. The failure surface may be planer along some well defined geological boundary, or may be arc-like. The size of relatively undeformed blocks range from 100 m^3 to 109 m^3. A planer slide is termed rock slide or debris slide depending on the material. Where the geologic materials are deep, uniform and cohesive, rotational slumps are common. A slump block does not usually move far but causes significant ground breakage.

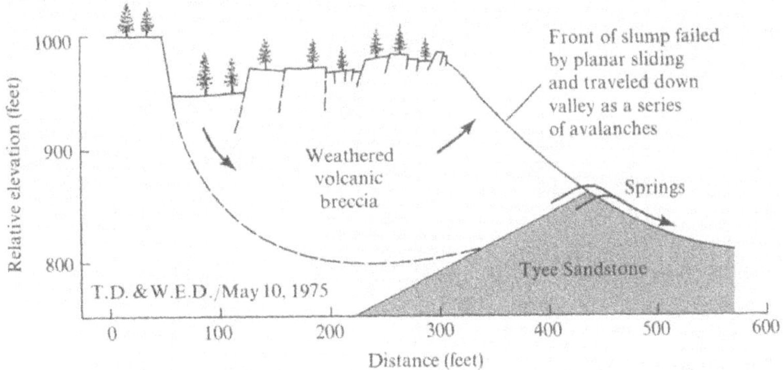

Cross section of Castle Creek Slump, Oregon

In flows the movement of debris resembles that of a fluid; more or less continuous internal deformation of the material. High proportion of air, water, and fine-grained materials favor flow. As the moving mass breaks into smaller blocks, the term changes from debris slide to debris avalanche to debris flow.

A small debris slide near the head of a valley can flow down the valley and develop into a huge debris flow as it incorporate water in the valley and scour the materials along the valley.

Earthflows consist mainly of fine-grained materials. They usually occur when a slump or other failure mobilizes clays that have a very high water content. Once in motion, the mass begins to flow and it develops a surging and spreading toe of dense, viscous material, which can flow over gentle gradients.

Mudflows develop where mass failures in fine-grained material mix with streams. The mudflows travel down canyons and spread over gently sloping fans of debris at the mountain front.

Soil creep is the slow downhill movement of debris that results from disturbance of soil by freezing and thawing, wetting and drying, or slow plastic deformation under the soil's own weight. It occurs on virtually all hillsides, and is responsible for most of the downslope transport of debris to streams channels in heavily vegetated areas.

Solifluction is the slow, viscous downslope flow of waterlogged soil underlain by an impervious layer, most commonly frozen ground. It is gaining considerable importance in development of arctic and subarctic regions.

Factors Controlling Mass Wasting

Planners do not need to know detailed engineering mechanics, but do need to know the nature of the forces controlling hillslope stability and the role of water in hillslope failures.

In this simple case of the soil block sliding along a the bedrock surface, W indicates weight of the block. The downslope component Wsinα is the driving force.

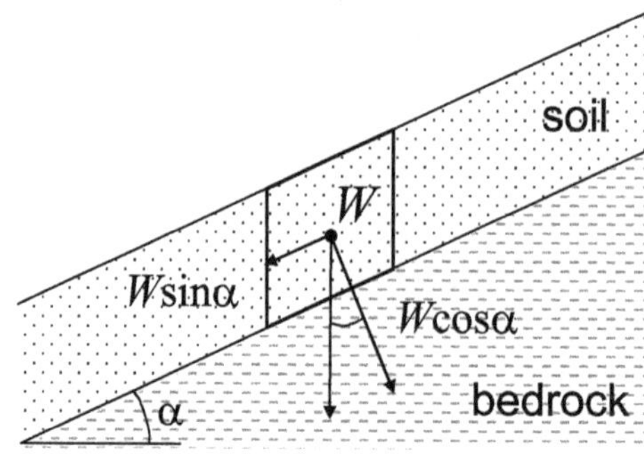

If Wsinα exceeds shear strength of the material, the slope failure may occur. The shear strength S is given by;

$$S = C_1 + C_2(W\cos\propto - P).$$

Where C_1 and C_2 are constants and P is the pressure of water. Noting that sinα increases and cosα decreases as α increases, failure is more likely in steeper hillslopes. Also note that S decreases as P increases. Large pressure of pore water weakens the material. This is why groundwater is so important in hillslope processes.

Suppose the water table rises during a storm. The pressure P at point A increases suddenly. This reduces the shear strength and causes a rotational failure.

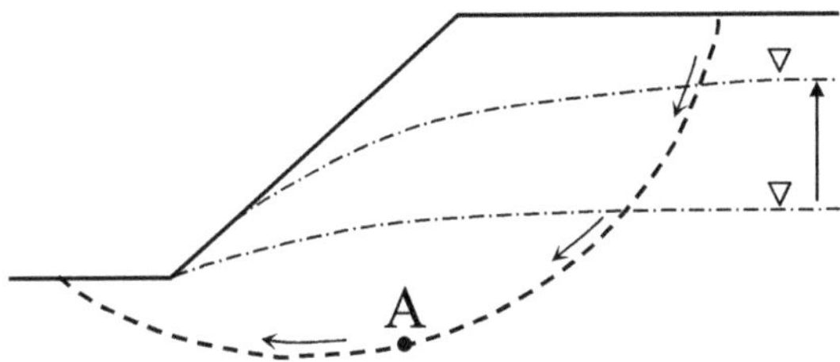

Geological Factors

Earth quakes can cause short-period increases in the downslope force and trigger landslides. In tectonically active areas, slow tilting of the ground by mountain building forces tends to increase slope gradients. The nature keeps adjusting gradients by occasional slope failures.

Human Factors

Undercutting during construction of highways and railroads increases the average slope gradients, and also increases the chance of slope failure.

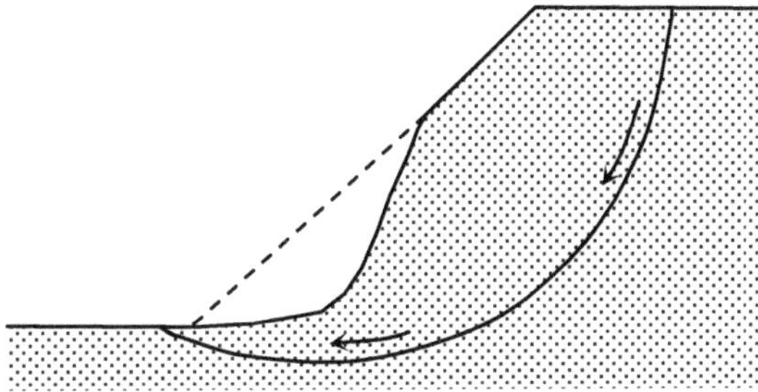

Overloading the top of hillslopes by housing construction is common. This extra weight may increase the chance of slope failure.

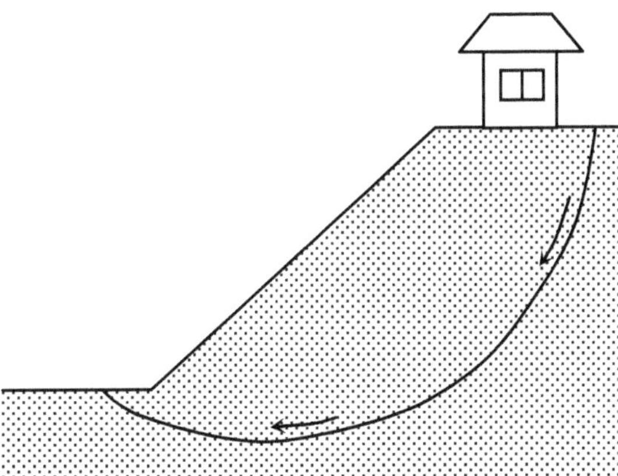

Altering the hydrology may have dramatic effects on hillslope stability. For example, clear cutting trees promotes soil erosion and weaken the support of soils by tree roots. It also reduces evapotranspiration and raise the water table.

Recognition and Avoidance of Landslide Hazards

Detailed analysis of hillslope stability requires the expertise of engineers. Planners need to decide;

1. Whether a site is in a stable area.
2. Whether there is enough uncertainty to warrant a detailed site investigation by an expert.
3. Whether the site is so obviously unstable that it should be avoided.

Techniques for evaluating landslide hazard involve recognizing at least one of the following.

1. Past hillslope failures.

   Direct evidence in airphoto, indirect evidence of altered vegetation, subtle topographic features, deposits formed by hillslope failures.

2. Conditions that are conductive to hillslope failures Steep slope gradients, mechanically weak geological material, poor permeability, high water table, seepage in the vicinity of steep slope.
3. De-stabilizing effects of planned development Undercutting, overloading, changing hydrologic conditions.

Igneous Processes

Landforms resulting from igneous processes may be related to eruptions of extrusive igneous rock material or emplacements of intrusive igneous rock. Volcanism refers to the extrusion of rock matter from Earth's subsurface to the exterior and the creation of surface terrain features as a result. Volcanoes are mountains or hills that form in this way. Plutonism refers to igneous processes that occur below Earth's surface including the cooling of magma to form intrusive igneous rocks and rock masses. Some masses of intrusive igneous rock are eventually exposed at Earth's surface where they comprise landforms of distinctive shapes and properties.

Volcanic Eruptions

Few spectacles in nature are as awesome as a volcanic eruption (figure below (a and (b)). Although large, violent eruptions tend to be infrequent events, they can devastate the surrounding environment and completely change the nearby terrain. Yet volcanic eruptions are natural processes and should not be unexpected by people who live in the vicinity of active volcanoes.

Eruptions can vary greatly in their size and character, and the volcanic landforms that result are extremely diverse. Explosive eruptions violently blast pieces of molten and solid rock into the air, whereas molten rock pours less violently onto the surface as flowing streams of lava in effusive eruptions. Variations in eruptive style and in the landforms produced by volcanism result mainly from temperature and chemical differences in the magma that feeds the eruption.

(a)

(b)

Figure above shown that: (a) Mount Vesuvius overlooks the ancient city center of Pompeii, near Naples, Italy. The eruption of Vesuvius in AD 79, which destroyed Pompeii, is an example of an episodic process. It is often difficult for humans to fully comprehend the potential danger from Earth processes that operate with bursts of intense activity, separated by years, decades, centuries, or even millennia of relative quiescence. (b) A plaster cast shows a victim who attempted to cover his face from hot gases and the volcanic ash that buried Pompeii.

Figure: Few natural events are as spectacular as a volcanic eruption. This eruption of Italy's Stromboli volcano, on an island off Sicily, lit up the night sky

Figure: Volcanic ash streaming to the southeast from Mount Etna on the Italian island of Sicily was captured on this photograph (south is at the top) taken from the International Space Station in July of 2001. The ash cloud reportedly reached a height of about 5200 meters (17,000 ft) on that day.

Located Under the Ash Cloud

The mineral composition that exists in a magma source is the most important factor determining the nature of a volcanic eruption. Silica-rich felsic magmas tend to be relatively cool in temperature while molten and have a viscous (thick, resistant to flowing) consistency. Mafic magmas are more likely to be extremely hot and less viscous, and thus flow readily in comparison to silica-rich magmas. Magmas contain large amounts of gases that remain dissolved when under high pressure at great depths. As molten rock rises closer to the surface, the pressure decreases, which tends to release expanding gases. If the gases trapped beneath the surface cannot be readily vented to the atmosphere or do not remain dissolved in the magma, explosive expansion of gases produces a violent, eruptive blast. Highly viscous, silica-rich magmas and lavas (rhyolitic in composition) have the potential to erupt with violent explosions. Mafic magmas and lavas, such as those with a basaltic composition, are hotter and less viscous (more fluid) and therefore tend to vent the gases more readily. When basaltic magma is forced to the surface, the resulting eruptions are usually effusive rather than explosive, and enormous amounts of fluid lava may be produced.

Molten material that is hotter, less viscous, and more mafic tends to erupt in the less violent effusive fashion, with streams of flowing lava. By contrast, the cooler, more viscous, silicic magma can produce explosive eruptions that hurl into the air molten material that solidifies in flight or on the surface or expel solid lava fragments of various sizes. These pyroclastic materials (from Greek: pyros, fire; clastus, broken), also referred to as tephra, vary in size from volcanic ash, which is sand-sized or smaller, to gravel-sized cinders (2–4 mm), lapilli (4–64 mm), and blocks (>64 mm). They may also include volcanic "bombs," which are large spindle-shaped clasts. In the most explosive eruptions, clay and silt-sized volcanic ash may be hurled into the atmosphere to an altitude of 10,000 meters (32,800 ft) or more (Figure above). The 1991 eruptions of Mount Pinatubo in the Philippines ejected a volcanic aerosol cloud that circled the globe. The suspended material caused spectacular reddish orange sunsets due to increased scattering and lowered global temperatures slightly for 3 years by increasing reflection of solar energy back to space.

Volcanic Landforms

The landforms that result from volcanic eruptions depend primarily on the explosiveness of the eruptions. We will consider six major kinds of volcanic landforms, beginning with those associated with the most effusive (least explosive) eruptions. Four of the six major landforms are types of volcanoes.

Lava Flows

Lava flows are layers of erupted rock matter that when molten poured or oozed over the landscape. After they cool and solidify they retain the appearance of having flowed. Lava flows can form from any lava type, but basalt is by far the most common because its hot eruptive temperature and low viscosity allow gases to escape, greatly reducing the potential for an explosive eruption. Basaltic lava flows may develop vertical fractures, called joints, due to shrinking of the lava during cooling. This creates columnar-jointed basalt flows (figure).

Figure: Basalt shrinks when it cools and solidifies. Some basaltic lava flows acquire a network of vertical cracks, called joints, upon cooling in order to accommodate the shrinkage. Often, polygonal joint systems separate vertical columns of basaltic rock creating columnar-jointed basalt as in this basalt flow in west-central Utah.

Lava flows display variable surface characteristics. Extremely fluid lavas can flow rapidly and for long distances before solidifying. In this case, a thin surface layer of lava in contact with the atmosphere solidifies, while the molten lava beneath continues to move, carrying the thin, hardened crust along and wrinkling it into a ropy surface form called pahoehoe. Lavas of slightly greater viscosity flow more slowly, allowing a thicker surface layer to harden while the still-molten interior lava keeps on flowing. This causes the thick layer of hardened crust to break up into sharp-edged, jagged blocks, making a surface known as aa. The terms pahoehoe and aa both originated in Hawaii, where effusive eruptions of basalt are common (figures below).

(a)

(b)

Figure: Scientists use Hawaiian terminology to refer to the two major surface textures commonly found on lava flows. Although all lava flows have low viscosity, slight variations exist from one flow to another. (a) Very low viscosity lava forms a ropy surface, called pahoehoe. (b) Somewhat more viscous lava leaves a blocky surface texture, called aa.

Lava flows do not have to emanate directly from volcanoes, but can pour out of deep fractures in the crust, called fissures, that can be independent of mountains or hills of volcanic origin. In some continental locations, very fluid basaltic lava that erupted from fissures was able to travel up to 150 kilometers (93 mi) before solidifying. These very extensive flows are often called flood basalts. In some regions, multiple layers of basalt flows have constructed relatively flat-topped, but elevated, tablelands known as basalt plateaus. In the geologic past, huge amounts of basalt have poured out of fissures in some regions, eventually burying existing landscapes under thousands of meters of lava flows. The Columbia Plateau in Washington, Oregon, and Idaho, covering 520,000 square kilometers (200,000 sq mi), is a major example of a basaltic plateau (figure below), as is the Deccan Plateau in India.

©Jeff Grass

Figure: River erosion has cut a deep canyon to expose the uppermost layers of basalt in the Columbia Plateau flood basalts in southwestern Idaho.

Shield Volcanoes

When numerous successive basaltic lava flows occur in a given region they can eventually pile up into the shape of a large mountain, called a shield volcano, which resembles a giant knight's shield resting on Earth's surface. The gently sloping, dome-shaped cones of Hawaii best illustrate this largest type of volcano. Shield volcanoes erupt extremely hot, mafic lava with temperatures of more than 1090°C (2000°F). Escape of gases and steam may hurl fountains of molten lava a few hundred meters into the air, with some buildup of cinders (fragments or lava clots that congeal in the air), but the major feature is the outpouring of fluid basaltic lava flows. Compared to other volcano types, these eruptions are not very explosive, although still potentially damaging and dangerous. The extremely hot and fluid basalt can flow long distances before solidifying, and the accumulation of flow layers develops broad, dome-shaped volcanoes with very gentle slopes. On the island of Hawaii, active shield volcanoes also erupt lava from fissures on their flanks so that living on the island's edges, away from the summit craters, does not guarantee safety from volcanic hazards. Neighborhoods in Hawaii have been destroyed or threatened by lava flows. The Hawaiian shield volcanoes form the largest volcanoes on Earth in terms of both their height— beginning at the ocean floor—and diameter.

Cinder Cones

The smallest type of volcano, typically only a couple of hundred meters high, is known as a cinder cone. Cinder cones generally consist largely of gravel-sized pyroclastics. Gas-charged eruptions throw molten lava and solid pyroclastic fragments into the air. Falling under the influence of gravity, these particles accumulate around the almost pipe like conduit for the eruption, the vent, in a large pile of tephra (figure below). Each eruptive burst ejects more pyroclastics that fall and cascade down the sides to build an internally layered volcanic cone. Cinder cone volcanoes typically have a rhyolitic composition, but can be made of basalt if conditions of temperature and viscosity keep gases from escaping easily. The form of a cinder cone is very distinctive, with steep straight sides and a large crater in the center, given the size of the volcano (figure below). Cinder cone examples include several in the Craters of the Moon area in Idaho, Capulin Mountain in New Mexico, and Sunset Crater, Arizona. In 1943 a remarkable cinder cone called Paricutín grew from a fissure in a Mexican cornfield to a height of 92 meters (300 ft) in 5 days and to more than 360 meters (1200 ft) in a year. Eventually, the volcano began erupting basaltic lava flows, which buried a nearby village except for the top of a church steeple.

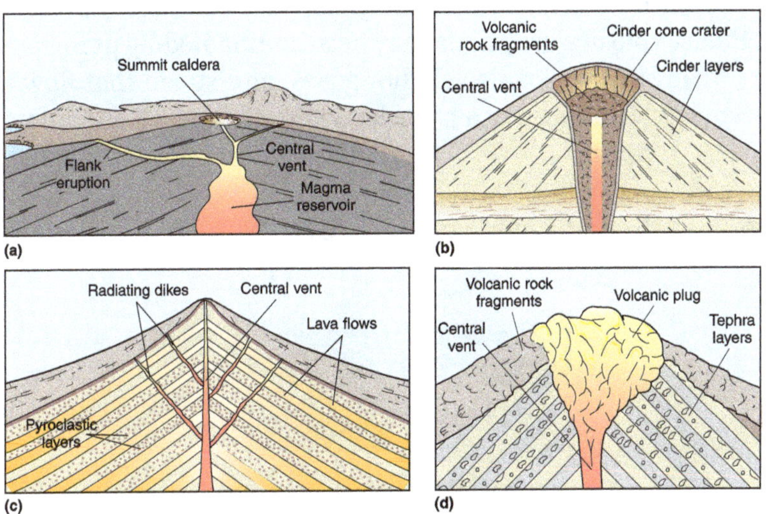

Figure: The four basic types of volcanoes are: (a) shield volcano, (b) cinder cone, (c) composite cone, also known as stratovolcano, and (d) plug dome.

Figure: Mauna Loa, on the island of Hawaii, is the largest volcano on Earth and clearly displays the dome, or convex, shape of a classic shield volcano. Mauna Loa reaches to 4170 meters (13,681 ft) above sea level, but its base lies far beneath sea level, creating almost 17 kilometers (56,000 ft) of relief from base to summit.

Composite Cones

A third kind of volcano, a composite cone, results when formative eruptions are sometimes effusive and sometimes explosive. Composite cones are therefore composed of a combination, that is, they represent a composite of lava flows and pyroclastic materials. They are also called stratovolcanoes because they are constructed of layers (strata) of pyroclastics and lava. The topographic profile of a composite cone represents what many people might consider the classic volcano shape, with concave slopes that are gentle near the base and steep near the top (Figure below). Composite volcanoes form from andesite, which is a volcanic rock intermediate in silica content and explosiveness between basalt and rhyolite. Although andesite is only intermediate in these characteristics, composite, cones are dangerous. As a composite cone grows larger, the vent eventually becomes plugged with unerupted andesitic rock. When this happens, the pressure driving an eruption can build to the point where either the plug is explosively forced out or the mountain side is pushed outward until it fails, allowing the great accumulation of pressure to be relieved in a lateral explosion. Such explosive eruptions may be accompanied by pyroclastic flows, fast-moving density currents of airborne volcanic ash, hot gases, and steam that flow downslope close to the ground like avalanches. The speed of a pyroclastic flow can reach 100 kilometers per hour (62 mi/hr) or more.

Hawaiian Volcanoes Erupt Less Explosively than Volcanoes of the Cascades or Andes

Figure: This fountain of lava in Hawaii reached a height of 300 meters (1000 ft).

Most of the world's best-known volcanoes are composite cones. Some examples include Fujiyama in Japan, Cotopaxi in Ecuador, Vesuvius and Etna in Italy, Mount Rainier in Washington, and Mount Shasta in California. The highest volcano on Earth, Nevados Ojos del Salado, is an andesitic composite cone that reaches an elevation of 6887 meters (22,595 ft) on the border between Chile and Argentina in the Andes, the mountain range after which andesite was named.

On May 18, 1980, residents of the American Pacific Northwest were stunned by the eruption of Mount St. Helens. Mount St. Helens, a composite cone in southwestern Washington that had been venting steam and ash for several weeks, exploded with incredible force on that day. A menacing bulge had been growing on the side of Mount St. Helens, and Earth scientists warned of a possible major eruption, but no one could forecast the magnitude or the exact timing of the blast. Within minutes, nearly 400 meters (1300 ft) of the mountain's north summit had disappeared by being blasted into the sky and down the mountainside (figure below). Unlike most volcanic eruptions, in which the eruptive force is directed vertically, much of the explosion blew pyroclastic debris laterally outward from the site of the bulge. An eruptive blast composed of an intensely hot cloud of steam, noxious gases, and volcanic ash burst outward at more than 320 kilometers per hour (200 mi/hr), obliterating forests, lakes, streams, and campsites for nearly 32 kilometers (20 mi). Volcanic ash and water from melted snow and ice formed huge mudflows that choked streams, buried valleys, and engulfed everything in their paths. More than 500 square kilometers (193 sq mi) of forests and recreational lands were destroyed. Hundreds of homes were buried or badly damaged. Choking ash several centimeters thick covered nearby cities, untold numbers of wildlife were killed, and more than 60 people lost their lives in the eruption. It was a minor event in Earth's history but a sharp reminder to the region's residents of the awesome power of natural forces.

Figure: Cinder cones grow as volcanic fragments (pyroclastics) ejected during gas-charged eruptions pile up around the eruptive vent. Here, a cinder cone stands among lava flows in Lassen Volcanic National Park, California.

Some of the worst natural disasters in history have occurred in the shadows of composite cones. Mount Vesuvius, in Italy, killed more than 20,000 people in the cities of Pompeii and Herculaneum in AD 79. Mount Etna, on the Italian island of Sicily, destroyed 14 cities in 1669, killing more than 20,000 people. Today, Mount Etna is active much of the time. The greatest volcanic eruption in recent history was the explosion of Krakatoa in the Dutch East Indies (now Indonesia) in 1883. The explosive eruption killed more than 36,000 people, many as a result of the subsequent tsunamis, large sets of ocean waves generated by a sudden offset of the water that swept the coasts of Java and Sumatra. In 1985 the Andean composite cone Nevado del Ruiz, in the center of Colombia's coffee growing region, erupted and melted its snowcap, sending torrents of mud and debris down its slopes to bury cities and villages, resulting in a death toll in excess of 23,000. The 1991 eruption of Mount Pinatubo in the Philippines killed more than 300 people and airborne ash caused climatic effects for 3 years following the eruption. In 1997 a series of violent eruptions from the Soufriere Hills volcano destroyed more than half of the Caribbean island of Montserrat with volcanic ash and pyroclastic flows. In recent years, Mexico City, one of the world's most populous urban areas, has been threatened by continued eruptions of Popocatepetl, a large, active composite cone that is 70 kilometers (45 mi) away. At this distance, ash falls from a major eruption would be the most severe hazard to be expected. Volcanic ash is much like tiny slivers of glass. It can cause breathing problems in people and other organisms. Vehicles stall when ash chokes the air intakes of combustion engines. In addition, the heavy weight of significant ash accumulations on roofs can cause buildings to collapse.

Figure: Composite cones are composed of both lava flows and pyroclastic material and have distinctive concave side slopes. Oregon's Mount Hood is a composite cone in the Cascade Range.

Plug Domes

Where extremely viscous silica-rich magma has pushed up into the vent of a volcanic cone without flowing beyond it, it creates a plug dome (figure). Solidified outer parts of the blockage create the dome-shaped summit, and jagged blocks that broke away from the plug or preexisting parts of the cone form the steep, sloping sides of the volcano.

(a)

(b)

(c)

Mount St. Helens, Washington, in the Cascade Range of the Pacific Northwest, illustrates the massive change that a composite volcano can undergo in a short period of time. (a) Prior to the 1980 eruption, Mount St. Helens towered majestically over Spirit Lake in the foreground. (b) On May 18, 1980, at 8:32 a.m., Mount St. Helens erupted violently. The massive landslide and blast removed more than 4.2 cubic kilometers (1 cu mi) of material from the mountain's north slope, leaving a crater more than 400 meters (1300 ft) deep. The blast cloud and monstrous mudflows destroyed the surrounding forests and lakes and took 60 human lives. (c) Two years after the 1980 eruption, the volcano continued to spew much smaller amounts of gas, steam, and ash. Mount St. Helens is currently experiencing a phase of eruptive activity that began in fall of 2004.

Great pressures can build up causing more blocks to break off, and creating the potential for extremely violent explosive eruptions, including pyroclastic flows. In 1903 Mount Pelée, a plug dome on the French West Indies island of Martinique, caused the deaths in a single blast of all but one person from a town of 30,000. Lassen Peak in California is a large plug dome that has been active in the last 100 years (figure). Other plug domes exist in Japan, Guatemala, the Caribbean, and the Aleutian Islands.

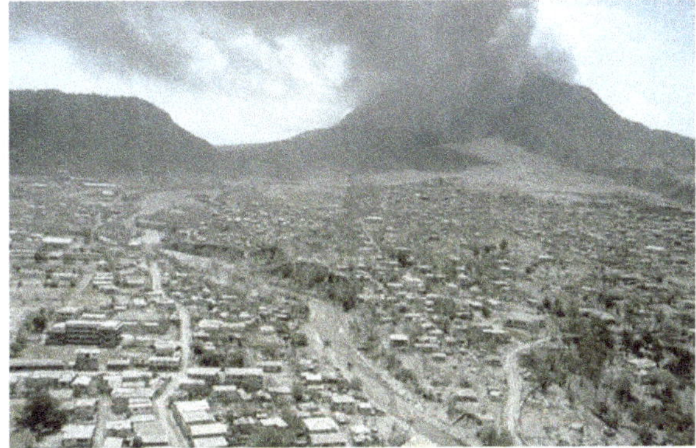

Figure: Beginning in 1995 the Caribbean island of Montserrat was struck by a series of volcanic eruptions, including pyroclastic flows, that devastated much of the island. The town of Plymouth, shown here, has been completely abandoned because of the amount of destruction and threat of future eruptions.

Plug Dome Volcanoes are Considered Dangerous

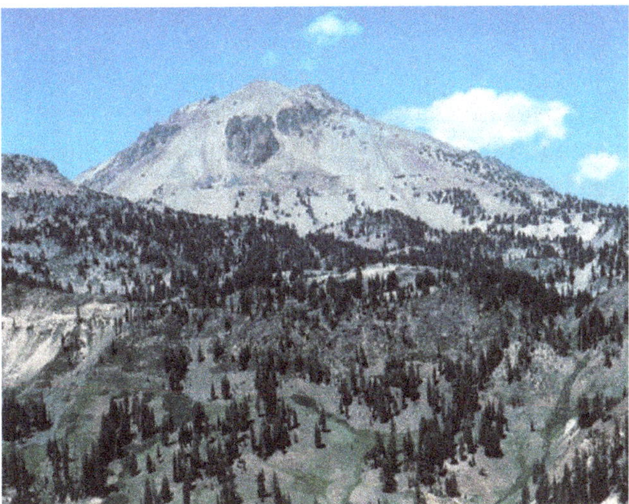

Figure: Plug dome volcanoes extrude stiff silica-rich lava and have steep slopes. Lassen Peak, located in northern California, is a plug dome and the southernmost volcano in the Cascade Range. The lava plugs are the darker areas protruding from the volcanic peak.

Figure: (a) Crater Lake, Oregon, is the best-known caldera in North America. It developed when a violent eruption of Mount Mazama about 6000 years ago blasted out solid and molten rock matter, (b) In the humid climate of south-central Oregon, water has accumulated in the crater, creating the 610-meterdeep (2000 ft) Crater Lake.

Calderas

Occasionally, the eruption of a volcano expels so much material and relieves so much pressure within the magma chamber that only a large and deep depression remains in the area that previously contained the volcano's summit. A large depression made in this way is termed a caldera. The best-known caldera in North America is the basin in south-central Oregon that contains Crater Lake, a circular body of water 10 kilometers (6 mi) across and almost 610 meters (2000 ft) deep, surrounded by near-vertical cliffs as much as 610 meters (2000 ft) high. The caldera that contains Crater Lake was formed by the prehistoric eruption and collapse of a composite volcano. A cinder cone, Wizard Island, has built up from the floor of the caldera and rises above the lake's surface (figure). The area of Yellowstone National Park is the site of three ancient calderas, and the Valles Caldera in New Mexico is another excellent example. Krakatoa in Indonesia and Santorini (Thera) in Greece have left island remnants of their calderas. Calderas are also found in the Philippines, the Azores, Japan, Nicaragua, Tanzania, and Italy, many of them occupied by deep lakes.

Plutonism and Intrusions

Bodies of magma that exist beneath the surface of Earth or masses of intrusive igneous rock that cooled and solidified beneath the surface are called igneous intrusions, or plutons. A great variety of shapes and sizes of magma bodies can result from intrusive igneous activity, also called plutonism. When they are first formed, smaller plutons have little or no effect on the surface terrain. Larger plutons, however, may be associated with uplift of the land surface under which they are intruded.

The many different kinds of intrusions are classified by their size, shape, and relationship to the surrounding rocks (figure). After millions of years of uplift and erosion of overlying rocks, even small intrusions may be located at the surface to become part of the landscape. Uplifted plutons composed of granite or other intrusive igneous rocks that are eventually exposed at the surface tend to stand higher than the landscape around them because their resistance to weathering and erosion exceeds that of many other kinds of rocks.

When exposed at Earth's surface, a relatively small, irregularly shaped intrusion is called a stock. A stock is usually limited in area to less than 100 square kilometers (40 sq mi). The largest intrusions, called batholiths when visible at the surface, are larger than 100 square kilometers and are complex masses of solidified magma, usually granite. Batholiths represent large plutons that melted, metamorphosed, or pushed aside other rocks as they developed kilometers beneath Earth's surface. Batholiths vary in size; some are as much as several hundred kilometers across and thousands of meters thick. They form the core of many major mountain ranges primarily because older covering rocks were eroded away, leaving the more resistant intrusive igneous rocks that comprise the batholith. The Sierra Nevada, Idaho, Rocky Mountain, Coast, and Baja California batholiths cover areas of hundreds of thousands of square kilometers of granite landscapes in western North America.

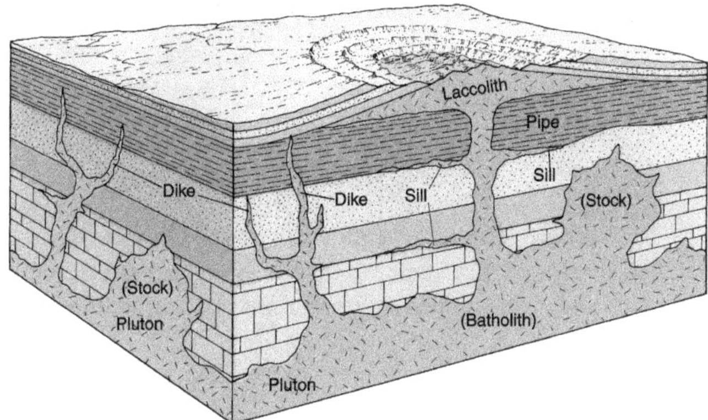

Figure: Igneous intrusions solidify below Earth's surface

Igneous intrusions solidify below Earth's surface. Because intrusive igneous rocks tend to be more resistant to erosion than sedimentary rocks, when they are eventually exposed at the surface sills, dikes, laccoliths, stocks, and batholiths generally stand higher than the surrounding rocks. Irregular, pod-shaped plutons less than 100 square kilometers (40 sq mi) in area form stocks when exposed, while larger ones form extensive batholiths

Magma can create other kinds of igneous intrusions by forcing its way into fractures and between rock layers without melting the surrounding rock. A laccolith develops when molten magma flows horizontally between rock layers, bulging the overlying layers upward, making a solidified mushroom-shaped structure. Laccoliths have a mushroom like shape because they are usually connected to a magma source by a pipe or stem. The resulting uplift on Earth's surface is like a giant blister, with magma beneath the overlying layers comparable to the fluid beneath the skin of a blister. Laccoliths are generally much smaller than batholiths, but both can form the core of mountains or hills after erosion has worn away the overlying less resistant rocks. The La Sal, Abajo, and Henry Mountains in southern Utah are composed of exposed laccoliths, as are other mountains in the American West.

Smaller but no less interesting landforms created by intrusive activity may also be exposed at the surface by erosion of the overlying rocks. Magma can intrude between rock layers without bulging them upward, solidifying into a horizontal sheet of intrusive igneous rock called a sill. The Palisades, along New York's Hudson River, provide an example of a sill made of gabbro, the intrusive compositional equivalent of basalt. Molten rock under pressure may also intrude into a non-horizontal fracture that cuts into the surrounding rocks. As it solidifies, the magma forms a wall-like structure of igneous rock known as a dike. When exposed by erosion, dikes often appear as vertical or near-vertical walls of resistant rock rising above the surrounding topography (figure). At Shiprock, in New Mexico, resistant dikes many kilometers long rise vertically to more than 90 meters (300 ft) above the surrounding plateau. Shiprock is a volcanic neck, a tall rock spire made of the exposed (formerly subsurface) pipe that fed a long-extinct volcano situated above it about 30 million years ago. Erosion has removed the volcanic cone, exposing the resistant dikes and neck that were once internal features of the volcano at Shiprock.

Figure: The La Sal Mountains in southern Utah, near Moab, are composed of a laccolith that was exposed at the surface by uplift and subsequent erosion of the overlying sedimentary rocks.

Figure: Sills develop where magma intrudes between parallel layers of surrounding rocks. The Palisades of the Hudson River, the impressive cliffs found along the river's western bank in the vicinity of New York City, are made from a thick sill of igneous rock that was intruded between layers of sedimentary rocks.

Figure: The igneous rock of this exposed dike in New Mexico was intruded into a near-vertical fracture in weaker sandstone. Later much of the sandstone was eroded away, leaving the resistant dike exposed.

Figure: Shiprock, New Mexico, is a volcanic neck exposed by erosion of surrounding rock. Volcanic necks are resistant remnants of the intrusive pipe of a volcano.

Tectonic Processes

The oft-used term tectonic processes is a grab-bag expression that encompasses all types of deformation, including the motion oftectonic plates, slip on individual faults, ductile deformation, and isostatic processes.

Tectonic is derived from τεχτονικη, which designated in ancient Greek the art of building. After its appearance in the geological literature of the mid-nineteenth century, it was mainly used to describe the study of Earth's structures on every scale, including those that are too large to be examined on the outcrop. It was only in the 1960s that mapping of the sea floor and mid-oceanic ridges conferred to the term tectonics its modern definition. Knowledge of the nature and age distribution of the oceanic crust provided a complete theory of Earth's dynamics, unifying earth sciences by pulling together diverse concepts and explaining many independent observations. The theory comprises four basic tenets.

a) The outer layer of Earth (its lithosphere) is relatively strong compared to the deeper asthenosphere; it is fragmented into pieces called plates.

b) The plates move independently relative to one another as rather rigid mechanical entities.

c) Earthquakes and volcanic eruptions are concentrated in narrow belts that correspond to linear topographic anomalies. This distribution of both the seismic and topographic disturbances indicates severe tectonic activity at or near the boundaries of the plates.

d) The interior of plates is relatively quiet, with far fewer earthquakes than occur at plate boundaries, and little volcanic activity.

Plate tectonics represents the simplest paradigm to relate superficial, geological, and geophysical structures with quantified movements attributed to deep processes of Earth. Plate tectonics permit in particular a dynamic interpretation of the present-day, large-scale features of Earth. Processes that opened oceanic basins through continents, and closed oceans to produce mountains and new continents, are collectively known as tectonic cycles. Since Earth is not expanding significantly, the rate of lithosphere destruction at convergent boundaries is virtually the same as the rate of creation at divergent boundaries.

Relative Plate Movements

Using the distribution of mountain ranges to define belts of continental shortening, and fold orientations to determine the force directions, large motions between continental blocks were envisioned by a few geologists who embraced the idea of continental drift, developed by the meteorologist Alfred Wegener in the early 1900s. Despite the work of some visionary scientists, geologists have not been able to document plate movements, because they needed to explore the oceans rather than the continents to find convincing arguments and demonstrate a mechanism for continental drift. This mechanism can be established for the past 200 million years, about the age of the oldest oceanic crust.

The lithosphere, which is the cool outer shell of Earth's convective system, includes both the crust and part of the upper mantle. The thickness of the lithosphere is thus controlled by the geotherm. It is defined on seismological criteria that reflect relatively gradual changes in various physical properties. As a metaphor, the lithospheric plates move over the asthenosphere a bit like ice-sheets on the sea. The asthenosphere is animated by the convection motions that drag the plates on the surface of the Earth, as fast as fingernails grow.

The edges of the plates are called plate boundaries. Points where three plates meet are triple junctions, which are succinctly classified in terms of the types of plate boundaries that join each other at the junction. Quadruple junctions are unstable configurations that immediately devolve into paired triple junctions.

The relative, horizontal movements of the rigid plates produce space problems at the contacts between adjacent plates. Because lithospheres are relatively rigid, their space problems produce forces that act on rocks. The response of the rocks is a deformation that creates secondary structures. These relative motions and their consequences are called plate tectonics.

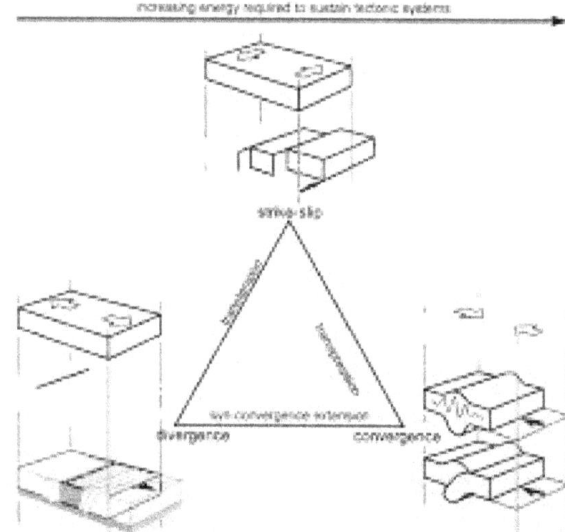

Figure . Ternary classification of plate movements and associated tectonic regimes

Horizontal velocities are up to nearly 20 cm yr -1, and relative plate motions are described in terms of convergence, divergence, and strike-slip.

- Divergence is due to adjoining plates moving away from one another. Crustal area extends, new ocean crust is formed at such boundaries, and plates grow.
- Convergence brings adjacent plates towards each other. The bulk crustal area diminishes, and plates even disappear down into the mantle along subduction zones. If Earth remains the same size, then the amount of crust consumed in a convergent zone must equal the amount of new crust that is formed in a divergent zone.
- Strike-slip consists of lateral shifting of one plate past another horizontally, without diverging or converging.

Plate movements may combine in many ways, depending on the kind of plate interaction that must be accommodated. The actual relative movement may be perpendicular or oblique to the plate boundary. An oblique convergence of plates produces transpressive deformation. An oblique divergence is termed transtension.

Any horizontal motion on a spherical Earth is necessarily a rotation about an axis through the center of the globe. Moving plates are rigid spherical caps that rotate around axes, each axis cutting Earth's surface at two antipodal points: the poles of rotation. The location of these poles depends on the reference used. Therefore, each pole pair corresponds to the relative rotation of one plate with respect to another, considered as fixed. These poles are the only points on the surface of Earth that do not move with respect to either of the two considered plates. Rotation poles are also called Euler poles, to honor the Swiss mathematician who established the basic geometry of movements on a sphere. The angular velocity between two plates is constant along the whole length of their boundary. Consequently, the linear velocity—that is, the relative displacement per unit time—increases away from the pole to accommodate larger and larger circumferences (figure below).

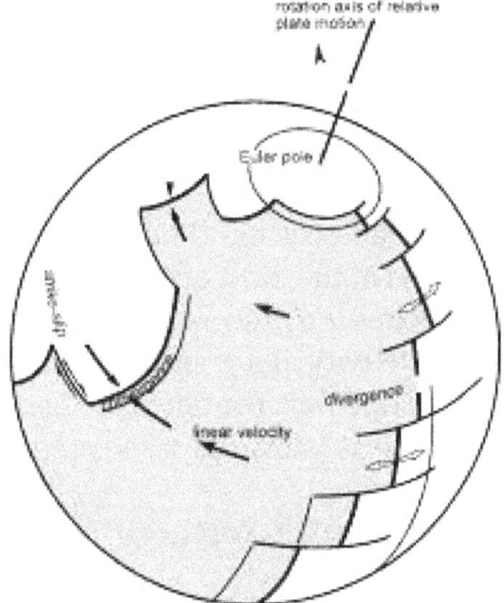

Figure: Description of a plate movement (shaded) with respect to an arbitrarily fixed plate (white) by an angle of rotation about a pole of rotation on a simplified two-plate planet, with one convergent, one divergent, and strike-slip boundaries. The angular rotation is the same everywhere, but the linear displacement increases according to the size of the sphere.

Divergent boundaries usually run along segments of great circles that intersect at the Euler poles. Therefore, divergent boundaries strike mostly orthogonal to the divergence direction. The orientation of convergent boundaries varies, because rigid shells on a sphere normally have arcuate boundaries. Ideal strike-slip boundaries are segments of small circles concentric about the Euler poles. Thus, they lie along the direction of relative motion between plates. A change of rotation pole between two plates requires a change in relative movement direction and, consequently, a change in the character of plate boundaries. Plate boundaries are not permanent features throughout their geological history. However, case studies suggest that poles for any two plates remain rather stable for long periods of times, pointing to some inertia in plate movements.

On the present Earth, the plate organization has formed two networks: one chain of divergent boundaries, and one chain of convergent boundaries, are segmented and connected by strike-slip boundaries (figure below). This simple organization reflects the mantle convection pattern, which is stable on a long term.

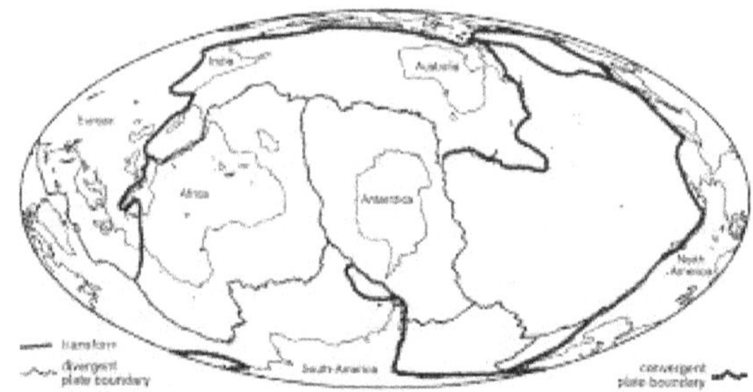

Figure: Framework of plate boundaries on Earth. Divergent and convergent boundaries pertain to two continuous belts connected though transform,strike-slip boundaries.

Plate Divergence

Divergent plate boundaries are zones of tension where plates split into two or more smaller plates that move apart, and the dominant stress field is extension. To accommodate the separation, dominantly normal faults and even open fissures form where crustal rocks undergo stretching, rupture, and lengthening coeval with lithosphere thinning. Because the lithosphere is thinned, there is upwelling of the mantle below the necked crust. Decompression of the mantle results in partial melting, and basaltic magma is injected into the fissures or extruded as fissure eruptions. Basaltic magmatism at the axis of a ridge creates new oceanic lithosphere as plates diverge from one another. Divergent plate boundaries are some of the most active volcanic areas on Earth. Magma filling the space opened between divergent plates is a process so important that more than half of Earth's surface has been created by volcanic activity along and within divergent boundaries during the past 200 million years. With sustained opening, continental edges move further and further from the mid-oceanic ridge. The whole process is known as seafloor spreading.

Rifts: Plate Divergence in Continental Settings

The break up of a continent is accomplished by normal faulting and produces rift systems. One of the best examples comprises the Read-Sea, the African Great Rift Valley, and the Gulf of Aden, which meet in a triple junction in the Afar region. Rifting produces long and linear crustal depressions whose characteristics are:

- An area where the crust has been arched upward.
- A relatively narrow width, which is about the thickness of the rifted continental crust, irrespective of the length of the rift valley.
- Steep margins that are as much as 3–4 km high normal-fault scarps. Some of these faults are very long, but most of them relay each other, eventually in an en echelon manner.
- Huge down-dropped blocks, which are sites of continental basins that are filled by thick (more than 1 km) clastic debris derived from adjacent high-standing blocks. Thinning of the continental crust has usually occurred on a series of listric faults.
- Parallel dyke swarms and outpouring flows of tholeiitic and alkaline basalts that accompany normal faults along which the crust is pulled apart. Mounts Kenya and Kilimanjaro are big volcanoes that exemplify this magmatism. Rhyolitic magma may be produced by partial melting of the granitic crust. The bimodal association of acid and basic volcanic rocks is characteristic of within-continent rift systems.

Continental rifts may represent the initial stages in the evolutionary cycle that further separates the older rifted segments, and finally leads to a continental break-up and ocean basin formation between two separated pieces of continental lithosphere. A rift that did not lead to continental separation remains preserved within the continent as a failed rift or aulacogen.

Two rifting modes have been envisioned, which refer to the role of the asthenosphere (figure below). The "mantle-activated" or "active" mechanism considers rifts to be initiated by mantle plumes or diapirs. The rising asthenosphere bends the lithosphere about a large dome, on top of which radial fractures bound rifts. The alternative "lithosphere activated" or "passive" mode attributes rifts to lithosphere extension under tectonic forces. These two modes are distinguished in their initial stage, plume-generated rifts beginning with doming and abundant volcanism, while passive rifts begin with narrow grabens of clastic sedimentation and younger, limited volcanism. In addition, active rifts tend to be symmetric above the vertically rising asthenosphere, whereas the mechanical response of a stretched lithosphere generates rather asymmetric systems. The East African rift has been considered to be typically mantle-activated, and the Baikal and Rhine Grabens to be lithosphere-activated. Passive rifts may evolve into active rifts by the upward intrusion of the asthenosphere into the necked lithosphere.

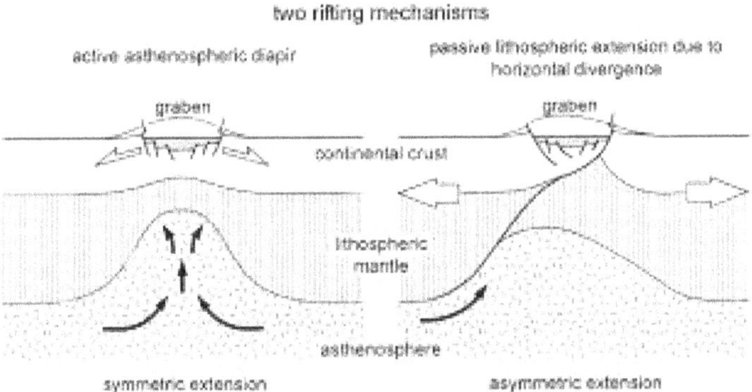

Figure: Extension mechanisms based on the active or passive role of the asthenosphere. Half arrows in the active mode point to outward plate movement imposed by the intruding asthenosphere. Large arrows in the passive mode represent far-field tectonic forces.

Passive Continental Margins

The natural evolution of a divergent plate boundary is recorded along the rifted, split, and spread apart margins of an original continent.

Stretching and thinning of the continental lithosphere at the margins of the two new continents results from initial rifting followed by extensive and continued normal faulting, often on listric faults in asymmetric systems. These concave, upward faults merge into or are cut by a flat detachment, on which extreme extension takes place. In the hanging wall of this master fault, sediments are generally deposited directly on tilted and eroded basement blocks, forming a profound unconformity. Synsedimentary faulting is common. Distance from the earth's surface to the top of the asthenosphere is steadily reduced as stretching proceeds. Alkaline igneous intrusions are common at that stage. Ultimately, the lithosphere is stretched and thinned to a point of rupture, permitting break-up of the continent, along with intrusion and extrusion of basaltic magma through the stretched lithosphere, creating new oceanic lithosphere. Oceanic lithosphere continues to crystallize in the developing separation zone between the adjacent diverging plates. Hence, a new and continually enlarging ocean basin forms at the site of the initial rift zone. Oceanic crust is welded directly onto continental crust. The mixture of original continental rocks and the added oceanic component at the continental edge produces a hybrid transitional crust. The thermal input is so important during the initial rifting stage that margins of the rifted continent are buoyantly uplifted.

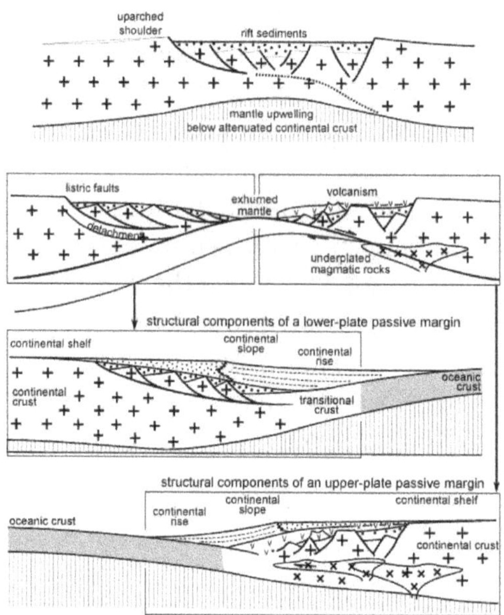

Figure: Structural components of extensional systems and idealized evolution from continental rifting to passive margin

The edge of the continent is a passive margin in that stresses are no longer deforming it (reference examples are the Atlantic margins). Structures and rock associations of the rifting phase are frozen into rocks of the continental margin, and they are passively conveyed laterally along with the rest of the plate to which they belong. The two continental margins become further away from the hot spreading center, and they cool down. Colder crust is denser, so the edges of the continental crust gradually subside below sea level; a non-mechanical process termed thermal subsidence. A pericontinental sea forms on the continental shelf. Shallow marine conditions may spread far over the continent to form an epicontinental sea (for example, the North Sea).

Passive margins include three main parts:

- A coastal plain and a submarine continental shelf of variable width (from a few kilometers—for example, Corsica—to over 1000 km—as in north-western Europe). They are generally underlain by a thick sequence of shallow-water mature clastic or biogenic sediments.
- The continental slope, at the edge of the continental shelf, which is generally present at the point where the shelf passes into a steeper topographic slope (3–5°) towards the basin.
- The continental rise, which links the ocean basin to the continental slope. A relatively thick sequence of sediments is generally present along the continental rise and slope.

Massive subsidence (up to 10–15 km) of the passive continental margin takes place as the attached oceanic lithosphere cools off. Most sediments on Earth lie on passive margins, which, therefore, contain more than half the world's oil reserves.

Ridges: Plate Divergence in Oceanic Settings

The present day divergent plate boundaries reside mainly in oceans where they form broad, fractured swells, generally more than 1000 km wide. The prominent physiographic expression is a world-encircling, approximately 100 km wide, and symmetrical topographic relief that rises up to 3000 m higher than the average ocean floor. It does so because the young lithosphere is hot and, therefore, less dense than the older and colder adjacent oceanic lithosphere. It is called the ocean ridge system. The dynamically active part of the system is restricted to a prominent axial rift valley that actually marks the plate boundary. Within a rift no wider than 20–30 km, the opening between diverging plates is continuously filled in by igneous intrusions of olivine tholeiite magma. New oceanic lithosphere is created by the combination of intrusion of mafic igneous rocks, extrusion of basaltic lavas interlayered with oceanic sediments, and extensional faulting. When cooled and crystallized, the intrusions and freshly accumulated mid-ocean-ridge-basalts (so called MORB) and sediments become part of the moving plates. They constitute new additions to the lithosphere. Accordingly, plate boundaries along oceanic ridges are also called constructive boundaries. As new crust forms, it continually spreads away from the ridge. This is termed accretion. The characteristics of a mid-oceanic ridge depend on its spreading rate. Slow ridges (such as the Atlantic) are higher and have more rugged topography than the fast ones (for example, the East Pacific Rise).

In response to changing tectonic conditions, a ridge may grow or propagate into an adjacent plate. Whole sections of a ridge may "jump" to form a new rift parallel to the existing ridge.

Marine Processes

Waves do much of the work of marine processes. They erode, transport and deposit materials. Waves are created by winds as they blow over the surface of the sea. It is the friction between the wind and water that sets waves in motion. The strength of waves depends on the strength of the wind depends on the strength of the wind. It also depends on the length of time and distance over which the wind has been blowing (the fetch).

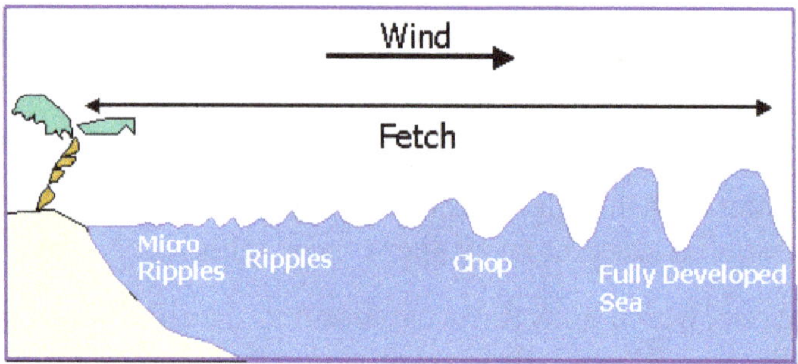

As waves near the coast, they enter into shallower water. Friction with the sea bed causes the wave to tip forward so that it eventually breaks. The resulting forward movement of the water, called the swash, runs up the beach until it runs out of energy. The water then runs back down the beach under gravity. This is called the backwash.

Constructive Waves

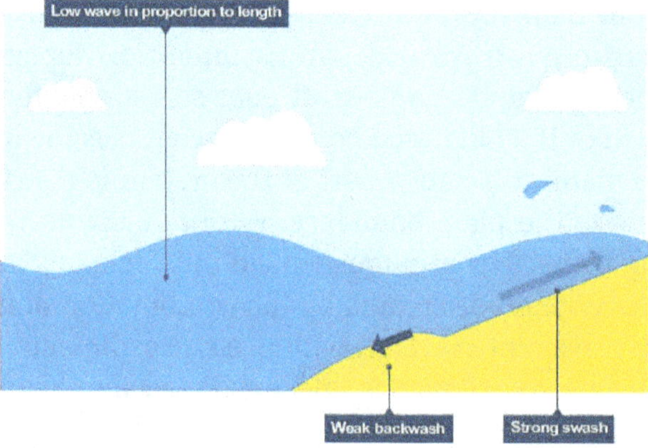

Destructive Waves

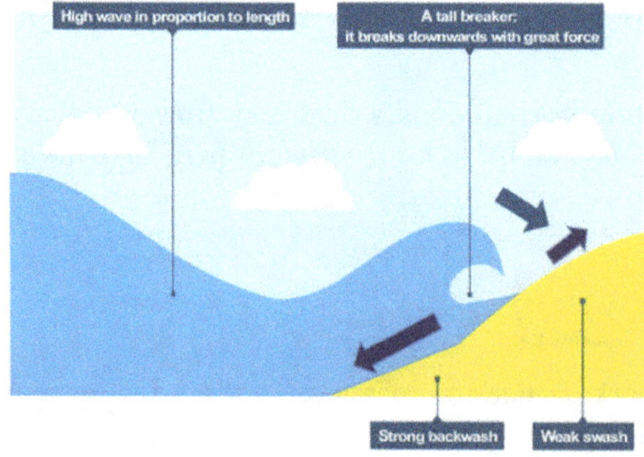

The balance between the swash and the backwash of waves created the difference between constructive and destructive waves. In constructive waves, the swash is stronger than the backwash. As a result, material is moved up the beach and much is left there (deposition). In destructive waves, the backwash is stronger. Material is dragged back down the beach (erosion) and moved along the coast by Long Shore Drift (LSD).

Processes of Erosion

It is destructive waves that do much of the erosion along a coast. They cut away at the coastline in a number of different ways:

- Hydraulic action – this results from the force of the waves hitting the cliffs and forcing pockets of air into cracks and crevices.
- Abrasion – this is caused by waves picking up stones and hurling them at cliffs and thus wearing the cliff away.
- Corrosion – the dissolving of rocks by the sea water.

Attrition is a process whereby the material carried by the waves becomes rounded and smaller over time as it collides with other material. It does not erode the coast as such but does form small pebbles and sand.

Long Shore Drift (LSD)

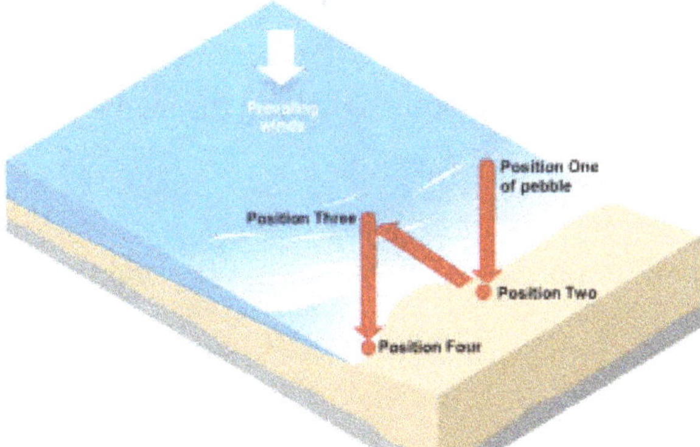

Once rocks are detached from the cliff, waves can move them along the coastline for quite long distances, this process is known as long shore drift. Generally speaking, the smaller the material, the further it is likely to be moved by waves as it is lighter. Eventually, the waves are unable to move so much material and the materials will be deposited to create new landforms.

Land Processes

There are three main processes at work on the landward side of the coastline:

- Weathering – the breakdown of rocks which is caused by freeze-thaw and the growth of salt crystals, by acid rain and by the growth of vegetation roots.
- Erosion – the weathering away of rocks by wind and rain.

- Mass Movement – the removal of cliff-face material under the influence of gravity in the form of rock falls, slumping and landslides.

Transportation

Transportation is the movement of sediment by the action of waves.

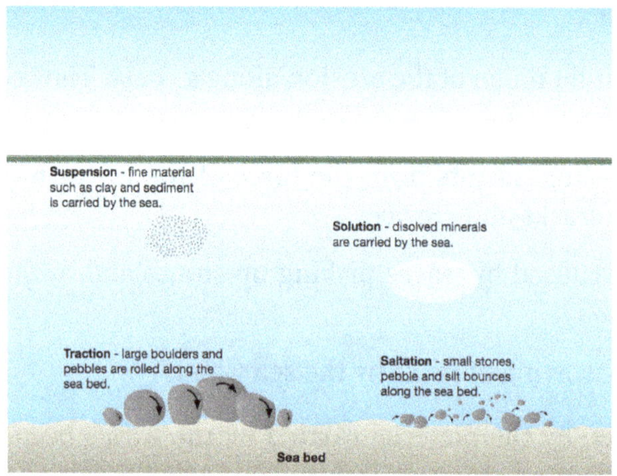

Traction

Traction involves the rolling of large and heavy rocks along the seabed.

Saltation

Saltation involves smaller material being bounced along the seabed.

Suspension

Suspension is when lighter sediment is suspended within the water. This often discolors the water close to the shore.

Solution

Sediment that has dissolved completely will be transported in solution.

Deposition

Deposition occurs when energy levels decrease in environments such as bays and estuaries. Where deposition occurs on the inside of a spit a salt marsh can form.

3

Weathering

Weathering refers to the slow breaking down of rocks, minerals, soil, wood, etc. due to to their contact with the Earth's water, atmosphere and organisms. There are two important classifications of weathering, physical and chemical weathering. Each of these can involve a biological influence. An elaborate study of the varied processes responsible for weathering, such as physical, biological and chemical weathering, has been explained in this chapter.

Weathering is the process of disintegration of rocks and soils and the minerals they contain through direct or indirect contact with the atmosphere. The weathering of an area occurs "without movement." By contrast, erosion involves the movement and disintegration of rocks and minerals by processes such as the flow of water, wind, or ice.

There are two main types of weathering: mechanical (or physical) and chemical. Mechanical weathering involves the breakdown of rocks and soils through direct contact with atmospheric conditions such as heat, water, ice, and pressure. Chemical weathering involves the direct effect of atmospheric chemicals or biologically produced chemicals (also called biological weathering). Chemical weathering alters the chemical composition of the parent material, but mechanical weathering does not. Yet, chemical and physical weatherings often go hand in hand. For example, cracks exploited by mechanical weathering will increase the surface area exposed to chemical action. Furthermore, the chemical action at minerals in cracks can assist the physical disintegration process.

The breakdown products following chemical weathering of rock and sediment minerals, and the leaching out of the more soluble parts, can be combined with decaying organic material to constitute soil. The mineral content of the soil is determined by the parent material (or bedrock) from which the minerals are derived. A soil derived from a single rock type is often deficient in one or more minerals for good fertility, while a soil weathered from a mix of rock types is often more fertile.

Weathering processes are of three main types: mechanical, organic and chemical weathering.

Mechanical or Physical Weathering

Mechanical weathering is also known as physical weathering. Mechanical weathering is the physical breakdown of rocks into smaller and smaller pieces. One of the most common mechanical actions is frost shattering. It happens when water enters the pores and cracks of rocks, then freezes. Frost weathering, frost wedging, ice wedging or cryofracturing is the collective name for several processes where ice is present. These processes include frost shattering, frost wedging and freeze-thaw weathering.

Once the frozen water is within the rocks, it expands by about 10% thereby opening the cracks a bit wider. The pressure acting within the rocks is estimated at 30,000 pounds per square inch at -7.6 °F. Over time, this pressure alongside the changes in weather makes the rock split off, and bigger rocks are broken into smaller fragments.

Another type of mechanical weathering is called salt wedging. Winds, water waves, and rain also have an effect on rocks as they are physical forces that wear away rock particles, particularly over long periods of time. These forces are equally categorized under mechanical or physical weathering because they release their pressures on the rocks directly and indirectly which causes the rocks to fracture and disintegrate.

Mechanical/physical weathering is also caused by thermal stress, which is the contraction and expansion effect on the rocks caused by changes in temperature. Due to uneven expansion and contraction, the rocks crack apart and disintegrate into smaller pieces.

Organic or Biological Weathering

Organic or biological weathering refers to the same thing. It is the disintegration of rocks as a result of the action by living organisms. Trees and other plants can wear away rocks since as they penetrate into the soil and as their roots get bigger, they exert pressure on rocks and make the cracks wider and deeper. Eventually, the plants break the rocks apart. Some plants also grow within the fissures in the rocks, which lead to widening of the fissures and then eventual disintegration.

Microscopic organisms like algae; moss, lichens and bacteria can grow on the surface of the rocks and produce chemicals that have the potential of breaking down the outer layer of the rock. They eat away the surface of the rocks. These microscopic organisms also bring about moist chemical microenvironments, which encourage the chemical and physical breakdown of the rock surfaces. The amount of biological activity depends upon how much life is in that area. Burrowing animals such as moles, squirrels and rabbits can speed up the development of fissures.

Chemical Weathering

Chemical weathering happens when rocks are worn away by chemical changes. The natural chemical reactions within the rocks change the composition of the rocks over time. Because the chemical processes are gradual and ongoing, the mineralogy of rocks changes over time thus making them wear away, dissolve, and disintegrate.

The chemical transformations occur when water and oxygen interacts with minerals within the rocks to create different chemical reactions and compounds through processes such as hydrolysis and oxidation. As a result, in the process of new material formations, pores and fissures are created in the rocks thus enhancing the disintegration forces.

Rainwater can also at times become acid when it mixes with acidic depositions in the atmosphere. Acid depositions are created in the atmosphere as a consequence of fossil fuel combustion that releases oxides of nitrogen, sulfur and carbon.

The resultant acid water from precipitation – (acid rain) reacts with the rock's mineral particles producing new minerals and salts that can readily dissolve or wear away the rock grains. Chemical weathering mostly depends on the rock type and temperature. For instance, limestone is more prone to chemical erosion compared to granite. Higher temperatures increase the rate of chemical weathering.

Physical Weathering

Physical weathering is also known as Mechanical weathering.

Mechanical weathering leads to the disintegration of rocks and wood. It usually produces smaller, angular fragments of material with the same properties as the original parent material (such as scree).

Thermal Expansion

Thermal expansion—also known as onion skin weathering, exfoliation, or thermal shock—is caused mainly by changes in temperature. It often occurs in hot areas such as deserts, where there is a large diurnal temperature range. The temperatures soar high in the day, while dipping to a few negative degrees at night. As the rock heats up and expands by day and cools and contracts by night, its outer layers undergo stress. As a result, the rock's outer layers peel off in thin sheets. Thermal expansion is enhanced by the presence of moisture.

Frost-induced Weathering

Frost-induced weathering, although often attributed to the expansion of freezing water captured in cracks, is generally independent of the water-to-ice expansion. It has long been known that moist soils expand (or "frost heave") upon freezing, as a result of the growth of ice lenses—water migrates from unfrozen areas via thin films to collect at growing ice lenses. This same phenomenon occurs within pore spaces of rocks. They grow larger as they attract water that has not frozen from the surrounding pores. The development of ice crystals weakens the rock, which, in time, breaks up.

Intermolecular forces between the mineral surfaces, ice, and water sustain these unfrozen films, which transport moisture and generate pressure between mineral surfaces as the lenses aggregate. Experiments show that porous rocks such as chalk, sandstone, and limestone do not fracture at the nominal freezing temperature of water of slightly below 0 °C, even when cycled or held at low temperatures for extended periods, as one would expect if weathering resulted from the expansion of water upon freezing. For the more porous types of rocks, the temperature range critical for rapid, ice-lens-induced fracture is -3 to -6 °C, significantly below freezing temperatures.

A rock in southern Iceland fragmented by freeze-thaw action.

Freeze-induced weathering action occurs mainly in environments where there is a lot of moisture, and temperatures frequently fluctuate above and below freezing point—that is, mainly alpine and periglacial areas. This process can be seen in Dartmoor, a southwest region of England, where it results in the formation of exposed granite hilltops, or tors.

Frost Wedging

Formerly believed to be the dominant mode, frost wedging may still be a factor in the weathering of nonporous rock, although recent research has demonstrated it less important than previously thought. Frost wedging—sometimes known as ice crystal growth, ice wedging, or freeze-thaw—occurs when water in the cracks and joints of rocks freezes and expands. In the expansion, it was argued that expanding water can exert pressures up to 21 megapascals (MPa) (2100 kilogram-force/cm^2) at −22 °C, and this pressure is often higher than the resistance of most rocks, causing the rock to shatter.

When water that has entered the joints freezes, the expanding ice strains the walls of the joints and causes the joints to deepen and widen. This is because the volume of water expands by about ten percent when it freezes.

When the ice thaws, water can flow further into the rock. Once the temperature drops below freezing and the water freezes again, the ice enlarges the joints further.

Repeated freeze-thaw action weakens the rocks, which eventually break up along the joints into angular pieces. The angular rock fragments gather at the foot of the slope to form a talus slope (or scree slope). The splitting of rocks along the joints into blocks is called block disintegration. The blocks of rocks that are detached are of various shapes, depending on their mineral structure.

Pressure Release

In pressure release (also known as unloading), overlying materials (not necessarily rocks) are removed by erosion or other processes, causing the underlying rocks to expand and fracture parallel to the surface. The overlying material is often heavy and the underlying rocks experience high pressure under it, such as in a moving glacier. Pressure release may also cause exfoliation to occur.

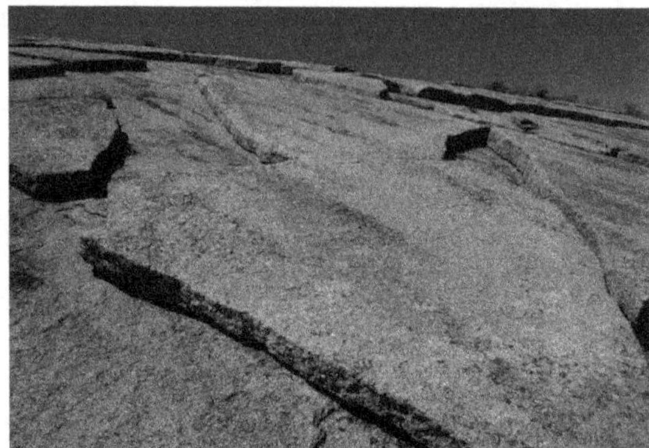

Pressure release of granite.

Intrusive igneous rocks (such as granite) are formed deep beneath the Earth's surface. They are under tremendous pressure because of the overlying rock material. When erosion removes the overlying rock material, these intrusive rocks are exposed and the pressure on them is released.

As a response to the decrease in pressure, the underlying rocks then expand upward. The expansion sets up stresses which cause fractures parallel to the rock surface to form. Over time, sheets of rock break away from the exposed rocks along the fractures. Pressure release is also known as "exfoliation" or "sheeting." These processes result in batholiths and granite domes, as found in Dartmoor.

Hydraulic Action

Hydraulic action refers to the action of water, generally from powerful waves, rushing into cracks in the rock face. This process traps a layer of air at the bottom of the crack, compressing it and weakening the rock. When the wave retreats, the trapped air is suddenly released with explosive force. The explosive release of highly pressurized air cracks away fragments at the rock face and widens the crack, so that more air is trapped on the next wave. This progressive system of positive feedback can damage cliffs and cause rapid weathering.

Salt-Crystal Growth

Salt crystallization, otherwise known as haloclasty, causes disintegration of rocks when saline solutions seep into cracks and joints in the rocks and evaporate, leaving salt crystals behind. These salt crystals expand as they are heated up, exerting pressure on the confining rock.

Salt crystallization may also take place when solutions decompose rocks. For example, limestone and chalk form salt solutions of sodium sulfate or sodium carbonate, of which the moisture evaporates to form their respective salt crystals.

Salts that have proved most effective in disintegrating rocks are sodium sulfate, magnesium sulfate, and calcium chloride. Some of these salts can expand up to three times or even more.

Weathering by salt crystallization is normally associated with arid climates, where strong heating causes rapid evaporation, leading to the formation of salt crystals. It is also common along coasts, and an example of salt weathering can be seen in the honeycombed stones in sea walls.

Salt weathering of building stone on the island of Gozo, Malta

Biotic Weathering

Living organisms may contribute to mechanical weathering as well as chemical weathering. Lichens and mosses grow on essentially bare rock surfaces and create a more humid chemical microenvironment. The attachment of these organisms to the rock surface enhances physical as well as chemical breakdown of the surface micro layer of the rock. On a larger scale, seedlings sprouting in a crevice and plant roots exert physical pressure and provide a pathway for water and chemical infiltration. Burrowing animals and insects disturb the soil layer adjacent to the bedrock surface, further increasing water and acid infiltration and exposure to oxidation processes.

Another well-known example of animal-caused biotic weathering is by the bivalve mollusc known as a Piddock. These animals, found boring into carboniferous rocks bore themselves further into the cliff-face.

Biological Weathering

Biological weathering is a type of weathering brought about by various activities of living organisms. Along with other types of weathering, biological weathering can contribute to the further degradation of rocks and rock particles by making them more susceptible to other environmental factors, whether be it biotic or abiotic factors.

Types of Biological Weathering

Living organisms can contribute to the process of weathering in many ways. Depending on the mechanism of how rocks and rock particles are broken down, biological weathering is of two types: by physical means or by chemicals and organic compounds.

Biological Weathering By Physical Means

This is a type of weathering that occurs when a force or pressure is applied to break rocks apart or degrade the minerals in them. By increasing the exposed surface area of rocks, they make it possible for other physical factors to speed up their degradation.

By Plants

- Plants can grow anywhere as long as there is water. Roots of trees or plants in general can biologically weather rocks by growing into the cracks and fractures of rocks and soil. As a result, they become more prone to breakage and eventually fall part.

By Animals

- Burrowing animals like shrews, moles, earthworms, and even ants contribute to biological weathering. In particular, these animals create holes on the ground by excavation and move the rock fragments to the surface. As a result, these fragments become more exposed to other environmental factors that can further enhance their weathering.

- When animals like birds forage for seeds and earthworms, they create holes and erode the upper surface of the soil, thus, contributes to weathering.

- An animal called the Piddock shell can drill into rocks in order for it to protect itself. By producing acids that can disintegrate the rock and turn it into fragments, it can create cracks and fractures and eat the minerals found in it.

- Like any other animal, humans can also indirectly contribute to biological weathering. By merely walking and running makes the soil particles crushed into smaller pieces. Other human activities such as planting and road construction can also contribute to biological weathering.

Biological Weathering By Chemicals/Organic Compounds

In this type of weathering, living organisms contribute through their organic compounds that contain molecules that acidify and corrode rock minerals. Because of such mechanism, biological weathering is also referred to as organic weathering.

By Plants

- When the roots of plants grow deeper into the soil, they tend to create cracks and crevices in marbles and limestones by producing certain acids that can eventually degrade them.

- According to studies, the mere presence of roots in the soil can wear out soil and rocks through the presence of humus. Humus, an organic component of the soil, can increase the availability of water, which then enhances the physical and chemical breakdown of rocks.

- When plants die, their roots (and other parts as well) are decomposed and are later on converted to organic matter, which produces carbon dioxide. This carbon dioxide (CO_2), when combined with water (H_2O), produces weak carbonic acid, which can degrade the surfaces of rocks and rock particles.

By Animals

- While ants and termites can contribute to the physical breakdown of rocks, these animals can also contribute to their biological degradation. Aside from creating holes and passages in the ground, these animals also make possible the easy passage of oxygen and water to the soil, which in turn, bring the dissolution of soil, rocks, and rock particles alike.

- When animals die, their bodies are converted to substances, which when combined with minerals found in the soil and rocks, can contribute to their degradation.

By Microorganisms

Despite their minute size, did you know that some microorganisms can also break down the largest of rocks and hardest of soil?

- One example of such activity is exhibited by lichens. In general, a lichen is a symbiosis between an algae and a fungus.

- The minerals in rocks are liberated when a fungus releases chemicals that can break them down. Such minerals are then consumed by the alga, further causing the wearing and development of cracks and gaps on the rock. As a result, cracked rocks become more prone to disintegration.

- Some fungi produce siderophores, a type of chelating agent, which can absorb various minerals and nutrients from the soil. By trading cations for hydrogen ions, siderophores make the soil more acidic, hence more prone to degradation.

- Another good example of biological weathering of rocks is by a group of bacteria called Actinomycetes. These bacteria through acid production, mineral solubilization, and metal leaching have successfully degraded rocks in Egypt.

Extreme Biological Weathering Examples

Ta Prohm – Ankor Wat Temple

Steeply Dipping Sedimentary Rock Strata

Sandstone in Lower Antelope Canyon, Arizona

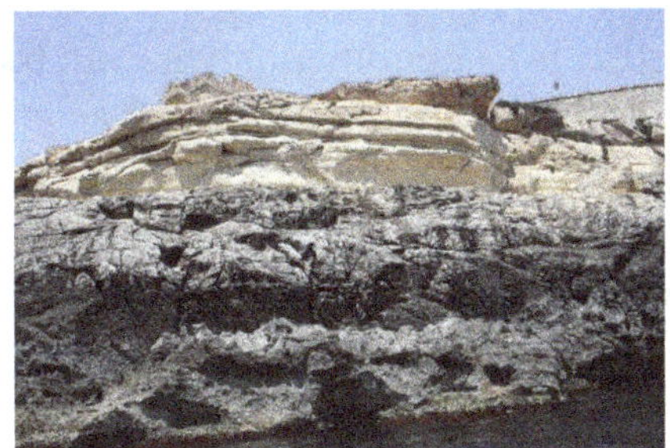

Sedimentary Rock With Sandstone

Advantages of Biological Weathering

Weathering is responsible for the creation of soil. Below are some other benefits of biological weathering:

Formation of Nutrient Rich Soil

The very process of weathering is what creates the soil, which then allows life to flourish on Earth. It is important to note that without it, minerals cannot be accumulated in the soil, hence, no nutrients will be available for plant and animal consumption.

Creation of Sediments

When broken, some rock pieces and particles eventually turn into sediments that are formed into different types of sedimentary rocks like sandstones and limestone. Usually, the broken rock pieces become deposited by rivers and are compacted by great pressure, enough to create a sedimentary rock.

Contribution to Land Formation

Like the formation of sediment rocks, biological weathering can also contribute to the formation of landmasses and landscapes. The process tends to be very slow, as it needs a lot of time to accumulate soil and rock particles, along with intense pressure to carry it out.

Disadvantages of Biological Weathering

While biological weathering has good effects, too much of it can be catastrophic and might pose a potential danger to life. The following are just some of them.

Erosion of Soil

Ironic how the very same process that creates soil for vegetation and makes possible the life on Earth is also the process that contributes to its erosion. When soil particles are broken down to smaller particles, it becomes a lot easier for it to be carried away by environmental factors like water and wind. In relation to this, when the topmost soil gets eroded, its fertility declines hence causing a simultaneous reduction in the productivity of the land.

Mass Wasting

Mass wasting is a phenomenon wherein rocks are broken down to smaller particles up to the point they make possible an erosion further enhanced by gravity. For instance, mudslide is a type of mass wasting wherein soil, and rock particles are mixed together forming a pool of mud that can bury almost anything. Another type of mass wasting are rockslides. As its name suggests, rock slides are huge masses of rocks fallen off by an avalanche.

Gradual Break Down

The saying "Life will always find a way" proves true for this example. Plant roots can grow in gaps beneath concrete and can create a force that is strong enough to break it. Sometimes, the effect can be damaging as it can make them more prone to breakage even with slight disturbances like an earthquake.

Chemical Weathering

Chemical weathering pertains to the changes in rock structure under the action or influence of chemical reactions. There are hundreds of natural chemical processes and reactions within the rocks the change the composition and the structure of the rocks over time. Temperature and, especially, moisture are critical for chemical weathering. Chemical weathering, therefore, occurs more quickly in hot, humid climatic regions.

When rain falls, the water is slightly acidic because carbon dioxide from the air dissolves in it. The rock may become weathered because of the minerals present in it that may react with the rainwater to form new minerals (clays) and soluble salts.

Chemical weathering happens because the processes are gradual and ongoing, therefore changing the mineralogy of the rocks over time that makes them to wear away, dissolve, or disintegrate. The rock's chemical transformations are highly influenced by the interaction of water and oxygen through processes such as hydrolysis and oxidation.

The ultimate end-result is the formation of new materials that contributes to the creation of pores and fissures in the rocks, in turn, accelerating the disintegration action. Chemical weathering involves various processes and types of weathering.

Processes of Chemical

Weathering

Storm water, acid rain, bio-chemical processes and mountain movements or rock uplifts are some of the processes determining chemical weathering.

Storm Water

Storm water plays an important role in the hydrolysis and oxidation processes within the rocks. Storm water can become a bit acidic by absorbing carbon dioxide in the atmosphere and as such this activates chemical action with the mineral granular particles in the rock producing chemical compounds such as salts and minerals that dissolve or eats away the rocks.

Acid Rain

Acid rain occurs when rainwater becomes acid by mixing with acidic depositions in the atmosphere. The combustion of fossil fuels such as coal, gasoline, and gas releases oxides of nitrogen, sulfur, and carbon into the air, which reacts with moisture to form rainwater that is more acidic than normal. The resulting acid rain then reacts with the rock's mineral particles to produce new minerals and salts that can easily dissolve or wear away the rock grains.

Bio-chemical Processes

Various plants and animals can create chemical weathering by emitting acidic compounds. As such, microscopic organisms such as moss, lichens, bacteria, and algae can speed up chemical weathering especially on the rock surfaces where they grow. They release what are termed as acidifying molecules (organic acids and protons) and chelating compounds (siderophores and organic acids). These compounds have the potential of breaking down iron and aluminum minerals in the rocks that are then dissolved by water, resulting in chemical weathering.

Rock Uplifts or Mountain Movement

The process of rock uplifts or mountain movements exposes new rocks to the atmospheric conditions such as moisture and extreme temperatures, allowing chemical weathering to take place. For instance, exposure of the rocks to surface waters or rainfall accelerates chemical weathering by hydrolysis and acidic reactions that eat away calcium ions and other minerals.

Types of Chemical Weathering

The various types of chemical weathering arise since it is driven by gradual and ongoing chemical reactions, mineralogy changes, the dissolving of the particles, and then the final wearing away or disintegration of the rocks. These reactions include hydrolysis, carbonation, dissolution, and oxidation. Temperature and most importantly moisture are vital for chemical weathering.

Hydrolysis

Hydrolysis is the chemical reactions caused by water. Water reacts with the rock and alters the size and chemical compositions of the minerals, lessening their resistance to weathering. Whenever minerals are hydrolyzed, crystal rocks and clay minerals such as calcium, potassium, and sodium ions are produced.

This type of chemical reaction is highly common in igneous rocks. The reaction takes either the form of hydration or dehydration. Absorption of water into the rock pertains to hydration while the removal of water from the rock pertains to dehydration. Hydration expands the rock's volume resulting in size alteration.

This is how gypsum is formed. Dehydration, on the other hand, reduces the volume of the rock. A good example is the formation of hematite from the removal of water from limestone.

Carbonation

Carbonation is the mixing of water with carbon dioxide to make carbonic acid. Carbonation takes place when the rock minerals react with weak carbonic acid formed when water combines with carbon dioxide in the atmosphere. Carbonic acid acts on the rock by breaking down and dissolving its mineral contents. The dissolved materials are washed away by ground water, and the soluble ions are stored in the groundwater supply. Rocks such as limestone and feldspar experience this type of chemical weathering more. This type of weathering is important in the formation of caves.

Dissolution

Dissolution equally means leaching. It the process by which the rocks are dissolved when exposed to rainwater. Limestone and rock salts are particularly the rocks that form solvent solutions when exposed to rainwater, surface waters, or even ground water. Upon dissolving, the minerals in the rocks become ion solutions in the water, which are then washed away. Karst features are a common example of this phenomenon.

Oxidation

Oxidation is another type of chemical weathering. Oxidation is also known as rusting. It is the process whereby the rock minerals lose one or more ions or atoms in the presence of oxygen. When minerals in the rock oxidize, they become less resistant to weathering. Oxygen combines with other substances via the oxidation process-giving rise to the ion or atom lose.

For instance, iron metal rusts because its ions change from one form to another by losing one electron. It becomes red or rust colored when oxidized. In a similar manner, iron-bearing minerals in rocks go through such a process by losing ions that alter its structure and size from one form to another. The wearing away of the rocks is thus sped up by oxidation/rusting as the resultant oxides are weaker than the original materials. Change of rock color is a prime example of rock disintegration by oxidation.

References

- Weathering: newworldencyclopedia.org, Retrieved 21 July 2018
- Different-types-of-weathering, geology: eartheclipse.com, Retrieved 16 May 2018
- Definition-processes-types-of-chemical-weathering, geology: eartheclipse.com, Retrieved 22 March 2018
- Biological-weathering: bioexplorer.net, Retrieved 12 April 2018

4

Aeolian, Glacial and Fluvial Processes and Landforms

Landforms are naturally-occurring features of the Earth. These are categorized according to the factors of elevation, slope, orientation, rock exposure, stratification, etc. This chapter closely examines some of the crucial geomorphological landforms, such as Aeolian, glacial, fluvial, coastal, volcanic and slope landforms.

Landforms are the natural features and shapes existent on the face of the earth. Landforms possess many different physical characteristics and are spread out throughout the planet. Together, landforms constitute a specific terrain and their physical arrangement in the landscape forms what is termed as topography. The physical features of landforms include slope, elevation, rock exposure, stratification and rock type.

Oceans and continents illustrate the largest grouping of landforms. They are they further subcategorized into many different landforms based on their physical features and shapes. Examples of distinctive landforms include mountains, valleys, plateaus, glaciers, hills, loess, deserts, shorelines, and plains. Features such as volcanoes, lakes, rivers, mid-ocean ridges, and the great ocean basins are also part of landform features.

Different Major Landforms on Earth

Major types of landforms on earth include mountains, valleys, plateaus, glaciers, hills, loess, plains and desserts.

Mountains

Mountains are lands physical features protruding high beyond the hills and very high up the land surface with steep top commonly shaped up to a peak. They are created through the action of incredible forces in the earth such as volcanic eruptions. Often, mountains occur in the ocean compared to land and some are seen as mountain islands as their peaks protrude out of the water. Mountain formation result from the forces of erosion, volcanism, or uplifts in the earth's crust.

The forces of heat and pressure within the earth's interior are the main influencing factors to these forces as stated by geologists. These forces can be summed up as the plate tectonic movements – theoretically defined as the division of the earth's outmost layer into several plates which are in constant motion. Hence, the uplifts are cause by collision or pulling apart of the plates that also triggers other various geologic activities such as the ejection of magma onto the surface or volcanic eruptions.

The movements also contribute to horizontal compression that is the deformation of crustal strata which gives rise to folds. The Himalayas and the Europe's Jura and Alps mountains are examples of mountains formed as a result of horizontal compression. Some mountain ranges are also formed as a result of wind, water, and ice erosion. Other mountains are created from volcanism.

Examples of volcanic mountains include Mount Fuji in Japan, Mount Vesuvius in Italy, Mount Erebus in Antarctica, and Mount Saint Helens in the United States. Majority of volcanic mountains have summit craters that still expel debris and steam.

Valleys

A valley is a lowland area or surface depression of the earth between higher lands such as mountains or hills. In simple terms, it can be defined as a natural trough bounded by mountains or hills on the surface of the earth sloping down to the lake, ocean or stream, which is created because of water or ice erosion. On this basis, the rivers or streams flowing through the valley empty the land's precipitation into the oceans.

The lowest parts of the valleys are very fertile and make very good farmlands. Majority of the valleys on land are made up of running streams and rivers and nearly all their floors slope downstream. Valleys within the mountains normally have narrow floors. The sides of a valley are termed as valley slopes or valley walls and the section of floor along riverbanks are referred to as flood plains.

Valleys physical features include U-shaped and V-shaped caused through the forces of erosion by the flowing masses that persistently widens and deepens the valley. The flowing masses are either water or glacier that carries away huge amounts of debris. Very narrow and deep valleys are known as canyons.

Plateaus

Plateaus are fairly flat areas higher than the land surrounding it. The surrounding areas may have very steep slopes. Some plateaus such as the Tibet are situated between mountain ranges. Plateaus cover wide land areas and together with their enclosed basins they cover approximately 45% of the entire earth's land surface.

Some plateaus, for instance the Columbia Plateau of the United States and the Deccan of India are basaltic and were created because of lava flows spreading to thousands of square kilometers thereby building up the fairly flat land surfaces. Other plateaus form as a result of upward folding while some are due to the erosion of the nearby land that leaves them elevated. Because plateaus are elevated, they are subject to erosion.

Low plateaus make up good farming regions whereas high plateaus are considered great for grazing livestock. Most of the world's high plateaus are deserts. Other typical examples of plateaus include the Bolivian plateau in South America, the Colorado plateau of the United States, the Laurentian Plateau and the plateaus of Iran, Arabia, and Anatolia.

Glaciers

Glaciers are the perennial ice sheets on the planet. They are huge masses of ice that slowly move over the land surface, predominant in high mountains and the cold Polar Regions. The very low temperatures in the regions are the enabling factor for the buildup of snow and densification into ice at depths of 15 meters or even more. Most glaciers have density thickness in the ranges of 91 to 3000 meters.

The movements begin when the compaction is so dense that it moves under the pressure of its weight. It is estimated that more than 75% of the world's fresh water is currently locked away in these frozen reservoirs. The glaciers include the Greenland Ice Sheet and the Antarctic Ice Sheet. The Antarctic Ice sheets outlet glaciers comprise the steep and extensively long and narrow depression Beordmore Glacier, which is one of the longest outlets in the world. The gradual rice in continental temperatures has seen the glacial density grow smaller owing to melting.

Hills

Hills are raised areas on the surface of the earth with distinctive summits, but are not as high as mountains. Hills are created as a result of accumulation of rock debris or sand deposited by wind and glaciers. They can also be created by faulting when the faults go slightly upwards. Hills are generally present in low mountain valleys and plains.

The Black Hills are the most known. Deep erosions of areas previously raised by the earth's crust disturbances carry most of the soil away leaving behind a hill. Human activities may also create hill when soils are dug and piled giant masses. Volcanic eruptions as well create hills after the eruption when the molten materials or lava cools and hardens in a pile.

Loess

Loess is a fine-grained unstratified accumulation of clay and silt deposited by the wind. It appears brown or yellowish in color and is brought about by past glacial activity in an area. In precise, it is sedimentary deposits of clay and silt mineral particles which take place on land in some parts of the world. The thickness of loess deposits are just a few meters and one of their basic feature is known as the 'cat steps'.

It's held together by few clay particles and is mostly composed of quartz crystals which readily slide against each other. This property makes it highly susceptible to erosion which leads to the 'cat steps' feature. Loess formed after the ice age when the glaciers covering a relatively large portion of the earth melted and was carried away, exposing the vast plains of mud.

Upon drying of the mud, the forces of wind blew away the mud and exposed sediments and eventually deposited them as silt in stacks on top of each other to create bold steep banks. Regions made of loess are witnessed in eastern China and the northwestern region of the United States.

Plains

Plains are broad flat areas on the earth's surface stretching over a wide area. Plains are lower than the land in their surrounding and can be found both inland and along the coast. Coastal plains rise from the seal level up to the point they meet raised landforms such as plateaus or mountains. The Atlantic Coastal plain is a prime example of a substantially populated and fertile coastal plain.

On the other hand, inland plains are generally found at high altitudes. Thick forests normally flourish on plains in humid climates. A fairly large portion of plains are covered by grasslands, for instance, the Great Plains in the United States. Human populations prefer settling on plains because of the soil and the terrain which is good for farming and building settlements such as cities, residential areas, and transportation networks. Flood plains are also in this category and they are formed as a result of continuous accumulation of sand, silt, and mud when rivers overflow its banks.

Deserts

Deserts are the hot and dry areas of the world. They are the arid and semi-arid lands with little or no vegetation. Deserts constitute approximately 20% of the earth's total land cover and are distinguished by little or no rainfall. The deserts are divided into four major categories including the Semi-Arid Deserts, the Hot and Dry Deserts, the Cold Deserts, and the Coastal Deserts.

These deserts are located in different areas of the world. Deserts experience very high temperatures, less cloud cover, low humidity, low atmospheric pressure, and very little rain, which makes them have very little vegetation cover. The soil cover is also rocky and shallow and with very little organic matter and as such, it only supports a few plants adapted to the conditions.

Plants such as cacti and short shrubs are the ones adapted to the desert conditions because they can conserve water and tolerate the high temperatures. Animals in the deserts include insects, small carnivores, snakes, lizards, and birds adapted to survive with very little water. These animals hide during the day till nightfall to avoid the heat. An example of a desert is the Sahara of North Africa.

Aeolian Landforms

Aeolian landforms are landforms created by the erosive and constructive effect of wind on the Earth's surface. The landforms are not only limited to the earth but have also been observed on other planets such as Mars where wind action is present. Wind transports and deposits sediments of various sizes in the areas where it is the chief agent of erosion. Such particles include silt, clay, and sand among others. Winds are mostly effective in areas where there is sparse vegetation, little or no soil moisture and unconsolidated elements. Such conditions are widely present in arid environments such as deserts. The Wind erodes the earth surface by both abrasion and deflation. Deflation refers to the removal of the loose particles on the surface of the Earth by the impact of wind. Abrasion is the wearing down of the surface of particles by the grinding action of the materials carried by the wind. The material eroded by winds are transported through saltation, suspension, and skipping on the earth's surface. The mode of transportation is largely dependent on their sizes and the strength of the wind. The deposition process by wind holds clues as to the past and present directions and intensities of wind. Wind-deposited bodies occur as sand sheets, ripples, and dunes.

Dune

In physical geography, a dune is a hill of sand formed by eolian (aeolian, or wind-related) processes. Dunes can take different forms and sizes, based on their interaction with the wind. Most kinds of dunes are longer on the windward side, where the sand is pushed up the dune, and are shorter on the "slip face" on the lee side (the side sheltered from the wind). The "valley," or trough, between dunes is called a slack. A dune field is an area covered by extensive sand dunes. Large dune fields are known as ergs.

Erg Awbari (Idehan Ubari) in the Sahara desert region of Fezzan in Libya.

Some coastal areas have one or more sets of dunes running parallel to the shoreline directly inland from the beach. In most cases, the dunes serve to protect the land against potential ravages by storm waves from the sea. Although the most widely distributed dunes are those associated with coastal regions, the largest complexes of dunes are found inland in dry regions and associated with ancient lake beds or sea beds.

Mesquite Flat Dunes in Death Valley National Park.

Dunes can also be formed under the action of water flow (alluvial processes), on sand or gravel beds of rivers, estuaries, and the seabed.

One of the main problems posed by sand dunes is their encroachment on human habitats. Sand dunes move by different means, all of them aided by the wind. Sand dunes threaten buildings and crops in Africa, the Middle East, and China. Preventing sand dunes from overwhelming cities and agricultural areas has become a priority for the United Nations Environment Program. On the other hand, dune habitats provide niches for highly specialized plants and animals, including numerous rare and endangered species. Some countries have developed extensive programs for dune protection.

Dune Shapes

Crescentic

Crescent-shaped mounds are generally wider than they are long. The slipface is on the dune's concave side. These dunes form under winds that blow from one direction, and they also are known as barchans, or transverse dunes. Some types of crescentic dunes move faster over desert surfaces than any other type of dune. A group of dunes moved more than 100 meters (m) per year between 1954 and 1959 in the People's Republic of China's Ningxia Province; similar rates have been recorded in the Western Desert of Egypt. The largest crescentic dunes on Earth, with mean crest-to-crest widths of more than 3 kilometers (km), are present in in China's Taklamakan Desert.

A diagram showing the formation of a dune with a slipface.

Linear

Straight or slightly sinuous sand ridges typically much longer than they are wide are known as linear dunes. They may be more than 160 kilometers long. Linear dunes may occur as isolated ridges, but they generally form sets of parallel ridges separated by miles of sand, gravel or rocky interdune corridors. Some linear dunes merge to form Y-shaped compound dunes. Many form in bidirectional wind regimes. The long axes of these dunes extend in the resultant direction of sand movement.

Linear loess hills known as pahas appear superficially similar. These hills appear to have been formed during the last ice age under permafrost conditions dominated by sparse tundra vegetation.

Star

Radially symmetrical, star dunes are pyramidal sand mounds with slipfaces on three or more arms that radiate from the high center of the mound. They tend to accumulate in areas with multidirectional wind regimes. Star dunes grow upward rather than laterally. They dominate the Grand Erg Oriental of the Sahara. In other deserts, they occur around the margins of the sand seas, particularly near topographic barriers. In the southeast Badain Jaran Desert of China, the star dunes are up to 500 meters tall and may be the tallest dunes on Earth.

Dome

Oval or circular mounds that generally lack a slipface, dome dunes are rare and occur at the far upwind margins of sand seas.

Parabolic

U-shaped mounds of sand with convex noses trailed by elongated arms are parabolic dunes. Sometimes these dunes are called U-shaped, blowout, or hairpin dunes, and they are well known in coastal deserts. Unlike crescent shaped dunes, their crests point upwind. The elongated arms of parabolic dunes follow rather than lead because they have been fixed by vegetation, while the bulk of the sand in the dune migrates forward.

Longitudinal (Seif) and Transverse Dunes

Longitudinal dunes (also called Seif dunes, after the Arabic word for "sword"), elongate parallel to the prevailing wind, possibly caused by a larger dune having its smaller sides blown away. Seif dunes are sharp-crested and are common in the Sahara. They range up to 300 m (900 ft) in height and 300 km (200 mi) in length. In the southern third of the Arabian Peninsula, a region called the Empty Quarter because of its total lack of population, a vast erg called Rub al Khali contains seif dunes that stretch for almost 200 km and reach heights of over 300 m.

Seif dunes.

Seif dunes are thought to develop from barchans if a change of wind direction occurs. The new wind direction will lead to the development of a new wing and the over development of one of the original wings. If the prevailing wind then becomes dominant for a lengthy period of time, the dune will revert to its barchan form, with one exaggerated wing. Should the strong wind then return the exaggerated wing will further extend so that eventually it will be supplied with sand when the prevailing wind returns. The wing will continue to grow under both wind conditions, thus producing a seif dune. On a seif dune, the slip face develops on the side facing away from the strong wind, while the slip face of a barchan faces the direction of movement. In the sheltered troughs between highly developed seif dunes, barchans may be formed because the wind is unidirectional.

A transverse dune is perpendicular to the prevailing wind, probably caused by a steady buildup of sand on an already existing minuscule mound.

Reversing Dunes

Occurring wherever winds periodically reverse direction, reversing dunes are varieties of any of the above shapes. These dunes typically have major and minor slipfaces oriented in opposite directions.

Complex dune: Dune 7 in the Namib desert, one of the tallest in the world.

All these dune shapes may occur in three forms: Simple, compound, and complex. Simple dunes are basic forms with a minimum number of slipfaces that define the geometric type. Compound dunes are large dunes on which smaller dunes of similar type and slipface orientation are superimposed, and complex dunes are combinations of two or more dune types. A crescentic dune with a star dune superimposed on its crest is the most common complex dune. Simple dunes represent a wind regime that has not changed in intensity or direction since the formation of the dune, while compound and complex dunes suggest that the intensity and direction of the wind has changed.

Dune Types

Sub-aqueous Dunes

Sub-aqueous (underwater) dunes form on a bed of sand or gravel under the actions of water flow. They are ubiquitous in natural channels such as rivers and estuaries, and also form in engineered canals and pipelines. Dunes move downstream as the upstream slope is eroded and the sediment deposited on the downstream or lee slope.

The Great Dune of Pyla is the largest dune in Europe.

These dunes most often form as a continuous "train" of dunes, showing remarkable similarity in wavelength and height.

Dunes on the bed of a channel significantly increase flow resistance, their presence and growth playing a major part in river flooding.

Lithified Dunes

A lithified (consolidated) sand dune is a type of sandstone that is formed when a marine or aeolian sand dune becomes compacted and hardened. Once in this form, water passing through the rock can carry and deposit minerals, which can alter the hue of the rock. Cross-bedded layers of stacks of lithified dunes can produce the cross-hatching patterns, such as those seen in Zion National Park.

A local slang term used for these consolidated dunes is "slickrock," a name used by pioneers of the old west because their steel-rimmed wagon wheels found it difficult to gain traction on this rock.

Coastal Dunes

Dunes form where "constructive waves" encourage the accumulation of sand, and where prevailing onshore winds blow this sand inland. There is a need for obstacles (for example, vegetation, pebbles) to trap the moving sand grains. As the sand grains get trapped they start to accumulate, this is the start of dune formation. The wind then starts to affect the mound of sand by eroding sand particles from the windward side and depositing them on the leeward side. Gradually this action causes the dune to "migrate" inland; as it does so it accumulates more and more sand. These dunes provide shelter from the wind.

Coastal dunes on the Kurnell Peninsula.

Ecological Succession on Coastal Dunes

As a dune forms, plant succession occurs. The conditions on an embryo dune are harsh, with salt spray from the sea carried on strong winds. The dune is well drained and often dry, and composed of calcium carbonate from seashells. Rotting seaweed, brought in by storm waves adds nutrients to allow pioneer species to colonize the dune. These pioneer species are marram grass, sea wort grass, and other sea grasses in the UK. These plants are well adapted to the harsh conditions of the fore-dune, typically having deep roots which reach the water table, root nodules that produce nitrogen compounds, and protected stoma, reducing transpiration. Also, the deep roots bind the sand together, and the dune grows into a fore dune as more sand is blown over the grasses. The grasses add nitrogen to the soil, meaning other, less hardy plants can then colonize the dunes. Typically these are heathers and gorses. These too are adapted to the low soil water content and have small, prickly leaves which reduce transpiration. Heathers add humus to the soil, but have a pH of lower than 7, so make the soil slightly acidic. Heathers are usually replaced by coniferous trees which can tolerate the low pH. Coniferous forests and heathland are common climax communities for sand dune systems.

Coastal dunes in Curonian spit.

Young dunes are called yellow dunes, dunes which have high humus content are called gray dunes. Leaching occurs on the dunes, washing humus into the slacks, and the slacks may be much more developed than the exposed tops of the dunes. It is usually in the slacks that more rare species are developed and there is a tendency for the dune slacks soil to be waterlogged and where only marsh plants can survive. These plants would include creeping willow, cotton grass, reeds, and rushes. Also, there is a tendency for natterjack toads to breed there.

Desertification

One of the biggest problems posed by sand dunes is their encroachment on human habitats. Sand dunes move through a few different means, all of them helped along by the wind. One way that dunes can move is through saltation, where sand particles skip along the ground like a rock thrown across a pond might skip across the water's surface. When these skipping particles land, they may knock into other particles and cause them to skip as well. With slightly stronger winds, particles collide in mid-air, causing sheet flows. In a major dust storm, dunes may move tens of meters through such sheet flows. Like snow, sand avalanches can fall down the steep slopes of the dunes facing away from the winds, also moving the dunes forward.

Sand threatens buildings and crops in Africa, the Middle East and China. Drenching sand dunes with oil stops their migration, but this approach is highly destructive to the dunes habitat and uses a finite resource. Sand fences might also work, but researchers are still analyzing optimum fence designs. Preventing sand dunes from overwhelming cities and agricultural areas has become a priority for the United Nations Environment Program.

One way of preventing sand from accumulating in roadways is planting trees and vegetation along the road.

Examples

- The dunes in the Thar desert in Rajasthan, India
- Tottori Sand Dunes,Tottori Prefecture, Japan
- The Kelso Dunes, in the Mojave desert of California
- Sands of Forvie within the Ythan Estuary complex, Aberdeenshire, Scotland
- Great Sand Dunes National Park, Colorado, United States
- Western Sahara

- White Sands National Monument, U.S.
- Rig-e Jenn in the Central Desert of Iran
- Rig-e Lut in the Southeast of Iran
- Indiana Dunes/Sleeping Bear Dunes National Lakeshore, Lake Michigan, U.S.
- Algodones Dunes near Brawley, California, U.S.
- Guadalupe-Nipomo Dunes, Central Coast California, U.S.
- Lencoi Maranhenses in the state of Maranhão, Brazil
- Mer'eb Dune (also written as Merheb) in United Arab Emirates, used as an arena for motor sports and skiing.
- Monahans Sandhills State Park near Odessa, Texas
- Oxwich Dunes, near Swansea on the Gower peninsula in Wales.
- Winterton Dunes-Norfolk, England
- The Killpecker sand dunes of Southwestern Wyoming, U.S.
- Jockey's Ridge State Park—Outer Banks, North Carolina

Conservation

Dune habitats provide niches for highly specialized plants and animals, including numerous rare and endangered species. Due to human population expansion dunes face destruction through recreation and land development, as well as alteration to prevent encroachment on inhabited areas. Some countries, notably the U.S., New Zealand, Great Britain, Australia, Canada, and the Netherlands have developed extensive programs of dune protection. In the UK, a Biodiversity Action Plan has been developed to assess dunes loss and prevent future dunes destruction.

Extraterrestrial Dunes

Dunes can likely be found in any environment where there is a substantial atmosphere, winds, and dust to be blown. Dunes are common on Mars, and have also been observed in the equatorial regions of Titan by the Cassini probe's radar.

Titan's dunes include large expanses with modal lengths of about 20-30 km. The regions are not topographically confined, resembling sand seas. These dunes are interpreted to be longitudinal dunes whose crests are oriented parallel to the dominant wind direction, which generally indicates west-to-east wind flow. The sand is likely composed of hydrocarbon particles, possibly with some water ice mixed in.

Loess

In some parts of the world, windblown dust and silt blanket the land. This layer of fine, mineral-rich material is called loess.

Loess is mostly created by wind, but can also be formed by glaciers. When glaciers grind rocks to a fine powder, loess can form. Streams carry the powder to the end of the glacier. This sediment becomes loess.

Loess ranges in thickness from a few centimeters to more than 91 meters (300 feet). Unlike other soils, loess is pale and loosely packed. It crumbles easily; in fact, the word "loess" comes from the German word for "loose." Loess is soft enough to carve, but strong enough to stand as sturdy walls. In parts of China, residents build cave-like dwellings in thick loess cliffs.

Extensive loess deposits are found in northern China, the Great Plains of North America, central Europe, and parts of Russia and Kazakhstan. The thickest loess deposits are near the Missouri River in the U.S. state of Iowa and along the Yellow River in China.

Loess accumulates, or builds up, at the edges of deserts. For example, as wind blows across the Gobi, a desert in Asia, it picks up and carries fine particles. These particles include sand crystals made of quartz or mica. It may also contain organic material, such as the dusty remains of skeletons from desert animals.

On the far side of the desert, moisture in the air causes the particles and dust to settle on the ground. There, grass and the roots of other plants trap the dust and hold it to the ground. More dust slowly accumulates, and loess is formed.

Loess often develops into extremely fertile agricultural soil. It is full of minerals and drains water very well. It is easily tilled, or broken up, for planting seeds. Loess usually erodes very slowly—Chinese farmers have been working the loess around the Yellow River for more than a thousand years.

The Loess Hills of western Iowa are more than 61 meters (200 feet) thick.

Properties

Loess is homogeneous, porous, friable, pale yellow or buff, slightly coherent, typically non-stratified and often calcareous. Loess grains are angular with little polishing or rounding and composed of crystals of quartz, feldspar, mica and other minerals. Loess can be described as a rich, dust-like soil.

Loess deposits may become very thick, more than a hundred meters in areas of China and tens of meters in parts of the Midwestern United States. It generally occurs as a blanket deposit that covers areas of hundreds of square kilometers and tens of meters thick.

Loess often stands in either steep or vertical faces. Because the grains are angular, loess will often stand in banks for many years without slumping. This soil has a characteristic called vertical cleavage which makes it easily excavated to form cave dwellings, a popular method of making human habitations in some parts of China. Loess will erode very readily.

In several areas of the world, loess ridges have formed that are aligned with the prevailing winds during the last glacial maximum. These are called "paha ridges" in America and "greda ridges" in Europe. The form of these loess dunes has been explained by a combination of wind and tundra conditions.

Loess near Hunyuan, Shanxi province, China.

Formation

According to Pye, four fundamental requirements are necessary for the formation of loess: a dust source, adequate wind energy to transport the dust, a suitable accumulation area, and a sufficient amount of time.

Periglacial Loess

Periglacial (glacial) loess is derived from the floodplains of glacial braided rivers that carried large volumes of glacial meltwater and sediments from the annual melting of continental icesheets and mountain icecaps during the spring and summer. During the autumn and winter, when melting of the icesheets and icecaps ceased, the flow of meltwater down these rivers either ceased or was greatly reduced. As a consequence, large parts of the formerly submerged and unvegetated floodplains of these braided rivers dried out and were exposed to the wind. Because these floodplains consist of sediment containing a high content of glacially ground flour-like silt and clay, they were highly susceptible to winnowing of their silts and clays by the wind. Once entrained by the wind, particles were then deposited downwind. The loess deposits found along both sides of the Mississippi River Alluvial Valley are a classic example of periglacial loess.

During the Quaternary, loess and loess-like sediments were formed in periglacial environments on mid-continental shield areas in Europe and Siberia, on the margins of high mountain ranges like in Tajikistan and on semi-arid margins of some lowland deserts like in China.

In England, periglacial loess is also known as brickearth.

Non-glacial

Non-glacial loess can originate from deserts, dune fields, playa lakes, and volcanic ash.

Some types of nonglacial loess are:

- Desert loess produced by aeolian attrition of quartz grains.
- Volcanic loess in Ecuador and Argentina.
- Tropical loess in Argentina, Brazil and Uruguay.
- Gypsum loess in Spain.
- Trade wind loess in Venezuela and Brazil.
- Anticyclonic loess in Argentina.

The thick Chinese loess deposits are non-glacial loess having been blown in from deserts in northern China. The loess covering the Great Plains of Nebraska, Kansas, and Colorado is considered to be non-glacial desert loess. Non-glacial desert loess is also found in Australia and Africa.

Fertility

Loess tends to develop into very rich soils. Under appropriate climatic conditions, it is some of the most agriculturally productive terrain in the world.

Soils underlain by loess tend to be excessively drained. The fine grains weather rapidly due to their large surface area, making soils derived from loess rich. One theory states that the fertility of loess soils is due largely to cation exchange capacity (the ability of plants to absorb nutrients from the soil) and porosity (the air-filled space in the soil). The fertility of loess is not due to organic matter content, which tends to be rather low, unlike tropical soils which derive their fertility almost wholly from organic matter.

Even well managed loess farmland can experience dramatic erosion of well over 2.5 kg /m^2 per year. In China the loess deposits which give the Yellow River its color have been farmed and have produced phenomenal yields for over one thousand years. Winds pick up loess particles, contributing to the Asian Dust pollution problem. The largest deposit of loess in the United States, the Loess Hills along the border of Iowa and Nebraska, has survived intensive farming and poor farming practices. For almost 150 years, this loess deposit was farmed with mouldboard ploughs and fall tilled, both intensely erosive. At times it suffered erosion rates of over 10 kilograms per square meter per year. Today this loess deposit is worked as low till or no till in all areas and is aggressively terraced.

Glacial Landforms

Glacial landforms are created by the action of the glacier through the movement of a large ice sheet. The glacial landforms can either be erosional or depositional depending on the action of the glacier. When glaciers retreat leaving behind crashed rocks and debris they create depositional landforms, but if the glaciers expand as a result of their accumulating weight crushing in the process and abrade scoured surface rock or bedrock, then it will lead to the formation of erosional landforms. Depositional landforms include eskers, kame, and Moraine while erosional landforms include Cirque, glacial horns, and arête. Apart from landforms, glaciers may also be striking features including lakes and ponds, particularly in the Polar Regions.

Valley

Valley is an elongate depression of the Earth's surface. Valleys are most commonly drained by rivers and may occur in a relatively flat plain or between ranges of hills or mountains. Those valleys produced by tectonic action are called rift valleys. Very narrow, deep valleys of similar appearance are called gorges. Both of these latter types are commonly cut in flat-lying strata but may occur in other geological situations.

Wherever sufficient rainfall occurs, opportunity exists for the land surface to evolve to the familiar patterns of hills and valleys. There are, of course, hyperarid environments where fluvial activity is minimal. There also are geomorphological settings where the permeability of rocks or sediments induce so much infiltration that water is unable to concentrate on the land surface. Moreover, some landscapes may be so young that insufficient time has elapsed for modification by fluvial action. The role of fluvial action on landscape, including long-term evolutionary processes, is considered here in detail.

Probably the world's deepest sub aerial valley is that of the Kāli Gandaki River in Nepal. Lying between two 8,000-metre (26,000-foot) Himalayan peaks, Dhaulāgiri and Annapūrna, the valley has a total relief of six kilometers (four miles). Because the Himalayas are one of the Earth's most active areas of tectonic uplift, this valley well illustrates the principle that the most rapid down cutting occurs in areas of the most rapid uplift. The reason for this seeming paradox lies in the energetics of the processes of degradation that characterize valley formation. The steeper the gradient or slope of a stream, the greater its expenditure of power on the streambed. Thus, as uplift creates higher relief and steeper slopes, rivers achieve greater power for erosion. As a consequence, the most rapid processes of relief reduction can occur in areas of most rapid relief production.

Perhaps the most famous example of a canyon is the Grand Canyon of the Colorado River in northern Arizona. The Grand Canyon is about 1.6 km (1 mile) deep and 180 meters (590 feet) to 30 km (19 miles) wide and occurs along a 443-km- (275-mile-) long reach where the Colorado River incised into a broad upwarp of sedimentary rocks.

Geomorphic Characteristics

The relief of valleys and canyons is produced by the incising action of rivers. Hillslope processes are indeed critical in the development of valley sides, but it is rivers that lower the level of erosion through degradation. Rivers ultimately adjust to a base level, defined as the lowest point at which potential energy can be transformed to the kinetic energy of river flow. In most cases, the ultimate base level for rivers is sea level. Some rivers drain to enclosed basins below sea level, as, for example, the Jordan River, which flows to the Dead Sea in Israel and Jordan. Moreover, rivers may adjust to local base levels, including zones of resistance to incision, lakes, and dams (both natural and artificial).

Valley Longitudinal Profiles

The longitudinal profile of a valley is the gradient throughout its length. Valleys formed by river action typically have a concave upward profile, steep in the headwaters and gentle in the lower reaches. The lower end of such a profile is adjusted to an effective lower limit of erosion defined by the base level.

In an ideal case of river adjustment to uniformly resistant materials, the longitudinal profile of a stream assumes a characteristic form that minimizes variations in transporting power. Power in a river derives from the rate of transfer of potential energy, dE/dt, which depends on the rate of fall in elevation of water, dy/dt, according to

$$\frac{dE}{dt} = mg\frac{dy}{dt},$$

where E is energy, t is time, m is mass, g is the acceleration of gravity, and y is elevation. The rate of fall in elevation, in turn, can be expressed as follows:

$$\frac{dy}{dt} = \frac{dy}{dx}\frac{dx}{dt} = SV$$

where S is the slope (fall in elevation, dy, with downstream horizontal distance, dx) and V is the flow velocity (change in horizontal distance, dx, with time, dt).

Combining equations (1) and (2) and using the fluid density ρ (mass per unit volume of water), one obtains

$$\frac{dE}{dt} = pg(W \cdot D \cdot L)SV,$$

where W is channel width, D is channel depth, L is a unit length of stream, and the other parameters are as defined above. Because flow discharge Q is defined as

$$Q = W \cdot D \cdot V,$$

the power per unit length of flow, Ω, can be expressed as

$$\Omega = dE / dt / L = pgQS.$$

It should be noted that in order to minimize variation in power, a river increasing its discharge in a downstream direction must decrease its slope. Thus, slope must be constantly decreasing downstream, explaining the concave upward character of the longitudinal profile.

The idealized concave upward longitudinal profile defined purely by energy considerations, only occurs where channel bed resistances and adequate adjustment time permit. Resistant zones of bedrock require greater power for a stream to incise at a given discharge Q than do less resistant zones. Therefore, by equation $\Omega = dE / dt / L = pgQS$. the stream gradient S must be locally steeper at resistant zones. Similarly, a rapid base-level change, such as a fall of sea level, may not allow adequate time for the entire longitudinal profile to adjust. One indication of such effects on a longitudinal profile is a nick point, or abrupt change in slope of the profile.

Valley Cross Profiles

The cross profiles of valleys involve a combination of fluvial and hillslope processes. Although slopes and rivers are often studied separately by process geomorphologists, hills and valleys are the features that dominate landscapes. In upland areas cross profiles of valleys are often narrow and deep. Canyon morphologies are most common. Further downstream, valley floors are wider and often dominated by floodplains and terraces.

Types of Valleys

One of the few classifications of valleys is that used by the German climatic geomorphologists Herbert Louis and Julius Büdel. In areas of rapid uplift and intense fluvial action such as tropical mountains, Kerbtal forms occur. These are characterized by steep, knife-edge ridges and valley slopes meeting in a V-shape. Where slopes are steep but a broad valley floor occurs, Sohlenkerbtal (meaning precisely a valley with such characteristics) is the prevailing form. Valleys of this kind develop under the influence of groundwater flow in Hawaii. Gutter-shaped valleys with convex sides and broad floors are called Kehltal; and broad, flat valleys of planation surfaces are termed Fachmuldental.

It is important to remember that the form of valleys reflects not only modern processes but also ancient ones. The entire valley or some landforms within it may be relict, with features inherited from past geologic periods during which occurred tectonic and climatic processes of intensities quite different from those prevailing today.

Hillslopes

Hillslopes constitute the flanks of valleys and the margins of eroding uplands. They are the major zones where rock and soil are loosened by weathering processes and then transported down gradient, often to a river channel.

Two major varieties of hillslopes occur in nature. On weathering-limited slopes, transport processes are so efficient that debris is removed more quickly than it can be generated by further weathering. Such hillslopes develop a faceted or angular morphology in which an upper free face, or cliff, contributes debris to a lower slope of accumulation. Slopes of this sort are especially common on bare rock where the profile of the slope is determined by the resistance of the rock, not by the erosional processes acting on it. One consequence of this is that many rock slopes retreat parallel to themselves in order to preserve the characteristic slope angle for a rock type of given strength. If the features of the rock change with depth into the slope, however, the characteristic angle of the slope will change. Rock slopes develop where weathering and soil erosion are slow (as in arid regions) and where rock resistance is high.

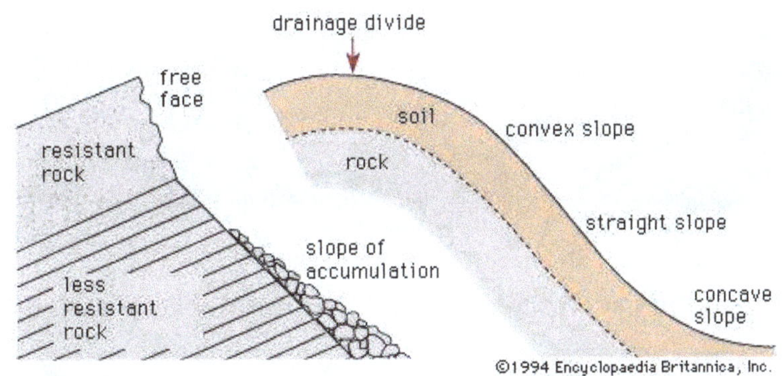

Comparison of idealized profiles for weathering-limited, faceted hillslopes (left) and transport-limited, sigmoid hillslopes (right).

The second major variety of slope is transport limited. Transport-limited slopes occur where weathering processes are efficient at producing debris but where transport processes are inefficient at removing it from the slope. Such slopes lack free faces and faceted appearances, and they are generally covered with a soil mantle. The profile of this type of slope generally has a sigmoid appearance, with convex, straight, and concave segments. The shape of the slope is an expression of the process acting upon it.

Convex slope segments commonly occur in the upper parts of soil-mantled slopes, as near the drainage divide. G.K. Gilbert elucidated the principles applying to convex slopes in his study of piles of mining-waste debris in California. The processes of soil creep and raindrop splash erode soil on the upper parts of slopes. Since soil eroded from the upper slope must pass each point below it, the volume of soil moved increases with distance from the divide. Since the transport rate for creep and rain splash is proportional to the slope angle, the slope angle must also increase from the divide, resulting in the slope convexity.

Straight slope segments are dominated by mass movement processes. Talus slopes are a type in which debris piles up to a characteristic angle of repose. When new debris is added to the slope, thereby locally increasing the angle, the slope adjusts by movement of the debris to reestablish the angle. Again, the result is a dynamic equilibrium in which the landform adjusts to processes acting upon it.

Concave slopes are especially common where overland-flow runoff transports sediment derived from upper slopes. Because the collection area for wash increases downslope and discharge Q is proportional to collection area, stream power—equation $\Omega = dE / dt / L = pgQS$.—can be maintained at lower slope angles. In addition, the size of particles being transported decreases downslope because of weathering and abrasion. Because the finer particles are easier to transport, slope angles can be reduced in the downslope direction. The result is a concave shape to the slope profile.

Erosional Landforms

Cirque/Corrie

Is an amphitheatre-shaped hollow basin cut into a mountain ridge. It has steep sided slope on three sides, an open end on one side and a flat bottom. When the ice melts, the cirque may develop into a tarn lake (figure).

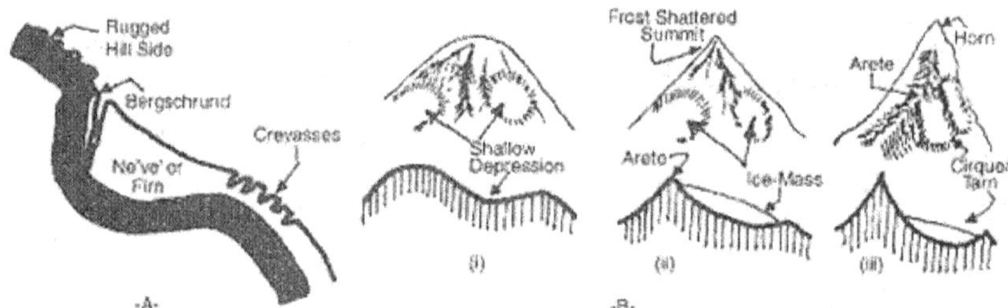

Figure: A. Different aspects of a cirque. B. Sequential development of cirque.

Glacial Trough

Glacial Trough is an original stream-cut valley, further modified by glacial action. Step- formation takes place at maturity; otherwise it is an ungraded and irregular feature.

'U' Shaped Valley

'U' Shaped Valley is another typically glacial feature. Since glacial mass is heavy and slow moving, erosional activity is uniform—horizontally as well as vertically. A steep sided and flat-bottomed valley results, which has a 'U' shaped profile.

Hanging Valley

Hanging Valley is formed when smaller tributaries are unable to cut as deeply as bigger ones and remain 'hanging' at higher levels than the main valley as discordant tributaries. This may happen due to glacial tilting or faulting.

Arete

Arete is a steep-sided, sharp-tipped summit with the glacial activity cutting into it from two sides.

Horn is a ridge that acquires a 'horn' shape when the piedmont glacier surrounds a summit.

Fjord is formed as a steep-sided narrow entrance-like feature at the coast where the stream meets the coast. Fjords are common in Norway, Greenland and New Zealand.

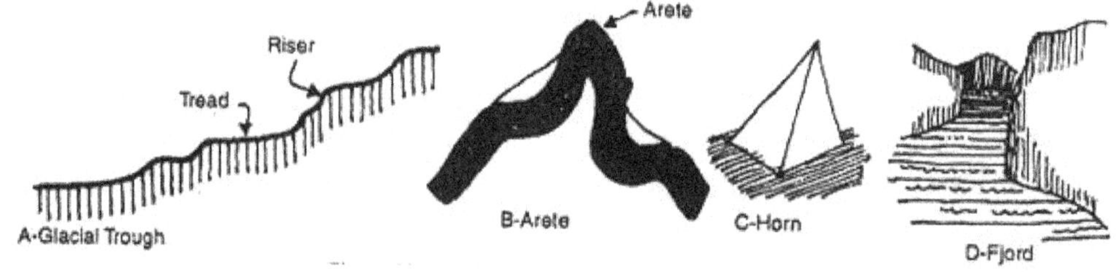

Figure: Landforme formed by glacial arosion

Mammillated Field

Mammillated Field is a term used for an ice-eroded field.

Depositional Landforms

Outwash Plain:

When the glacier reaches its lowest point and melts, it leaves behind a stratified deposition material, consisting of rock debris, clay, sand, gravel etc. This layered surface is called till plain or an outwash plain and a downward extension of the deposited clay material is called valley train (figure).

Esker:

Is a winding ridge of unassorted depositions of rock, gravel, clay etc. running along a glacier in a till plain. The eskers resemble the features of an embankment and are often used for making roads. If the melting of glacier has been punctuated, it is reflected in a local widening of the esker and here it is called a beaded esker (figure).

Kame:

Terraces are the broken ridges or unassorted depositions looking like hummocks in a till plain (figure).

Drumlin:

Is an inverted boat-shaped deposition in a till plain caused by deposition. The erosional counterpart is called a roche moutonne (figure).

Kettle Holes:

Can be formed when the deposited material in a till plain gets depressed locally and forms a basin (figure).

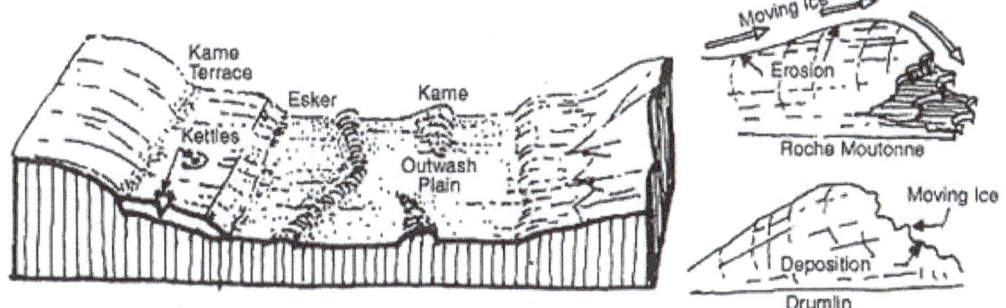

Figure: Landforme formed by glacial deposition

Moraine:

Moraine is a general term applied to rock fragments, gravel, sand, etc., carried by a glacier. Depending on its position, the moraine can be ground, lateral, medial or terminal moraine. The material dropped at the end of a valley glacier in the form of a ridge is called the terminal moraine. Each time a glacier retreats, a fresh terminal moraine is left at a short distance behind the first one.

The material deposited at either of its sides is known as lateral moraine. When two glaciers join, their lateral moraines also join near their confluence and are called medial moraines. Many Alpine pastures in the Himalayas like the Margs of Kashmir occupy the sites of morainic deposits of old river valleys. The excessive load that cannot be carried forward by a glacier is deposited on its own bed or at the base and appears as what is known as ground moraine. (figure below)

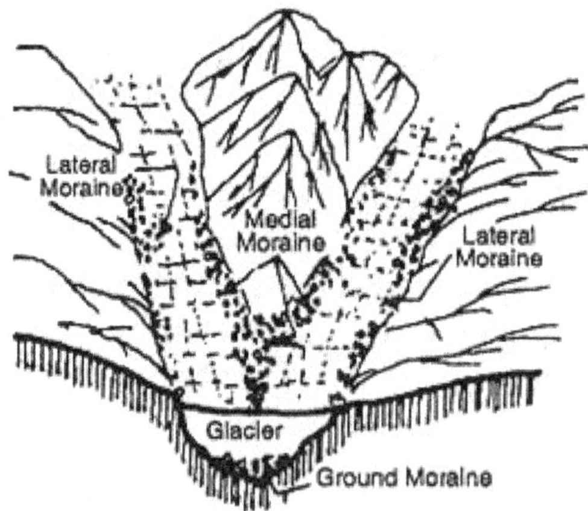

Figure: Different positions of moraines

Glacial Cycle of Erosion

Youth:

The stage is marked by the inward cutting activity of ice in a cirque. Aretes and horns are emerging. The hanging valleys are not prominent at this stage.

Maturity:

The valley glacier gets transformed into trunk glacier and hanging valleys start emerging. The opposite cirques come closer and the glacial trough acquires a stepped profile which is regular and graded.

Old Age:

Emergence of a 'U'-shaped valley marks the beginning of old age. An outwash plain with features such as eskers, kame terraces, drumlins, kettle holes etc. is a prominent development. The opposite cirques coalesce and the summit heights are greatly reduced. Mountain tops become rounded.

Other Types of Glacial Landforms

Mesa

A mesa is a flat-topped mountain or hill. It is a wide, flat, elevated landform with steep sides.

Mesa is a Spanish word that means table. Spanish explorers of the American southwest, where many mesas are found, used the word because the tops of mesas look like the tops of tables.

Mesas are formed by erosion, when water washes smaller and softer types of rocks away from the top of a hill. The strong, durable rock that remains on top of a mesa is called cap rock. A mesa is usually wider than it is tall.

Mesas are usually found in dry regions where rock layers are horizontal. The Grand Mesa in the U.S. state of Colorado, considered the largest mesa in the world, has an area of about 1,300 square kilometers (500 square miles) and stretches for 64 kilometers (40 miles).

This gorgeous mesa towers above the surrounding vegetation of the Guiana Highlands of South America.

Formation

Har Qatum, a mesa located on the southern edge of Makhtesh Ramon, Israel

Mesas form by weathering and erosion of horizontally layered rocks that have been uplifted by tectonic activity. Variations in the ability of different types of rock to resist weathering and erosion cause the weaker types of rocks to be eroded away, leaving the more resistant types of rocks topographically higher than their surroundings. This process is called differential erosion. The most resistant rock types include sandstone, conglomerate, quartzite, basalt, chert, limestone, lava flows and sills. Lava flows and sills, in particular, are very resistant to weathering and erosion, and often form the flattop, or caprock, of a mesa. The less resistant rock layers are mainly made up of shale, a softer rock that weathers and erodes more easily.

The differences in strength of various rock layers is what gives mesas their distinctive shape. Less resistant rocks are eroded away on the surface into valleys, where they collect water drainage from the surrounding area, while the more resistant layers are left standing out. A large area of very resistant rock, such as a sill may shield the layers below it from erosion while the softer rock surrounding it is eroded into valleys, thus forming a caprock.

Differences in rock type also reflect on the sides of a mesa, as instead of smooth slopes, the sides are broken into a staircase pattern called "cliff-and-bench topography". The more resistant layers form the cliffs, or stairsteps, while the less resistant layers form gentle slopes, or benches, between the cliffs. Cliffs retreat and are eventually cut off from the main cliff, or plateau, by basal sapping. When the cliff edge does not retreat uniformly, but instead is indented by headward eroding streams, a section can be cut off from the main cliff, forming a mesa.

Basal sapping occurs as water flowing around the rock layers of the mesa erodes the underlying soft shale layers, either as surface runoff from the mesa top or from groundwater moving through permeable overlying layers, which leads to slumping and flowage of the shale. As the underlying shale erodes away, it can no longer support the overlying cliff layers, which collapse and retreat. When the caprock has caved away to the point where only a little remains, it is known as a butte.

Canyon

A canyon is a deep, narrow valley with steep sides. "Canyon" comes from the Spanish word canon, which means "tube" or "pipe." The term "gorge" is often used to mean "canyon," but a gorge is almost always steeper and narrower than a canyon.

The movement of rivers, the processes of weathering and erosion, and tectonic activity create canyons.

River Canyons

The most familiar type of canyon is probably the river canyon. The water pressure of a river can cut deep into a river bed. Sediments from the river bed are carried downstream, creating a deep, narrow channel.

Rivers that lie at the bottom of deep canyons are known as entrenched rivers. They are entrenched because, unlike rivers in wide, flat flood plains, they do not meander and change their course.

The Yarlung Zangbo Grand Canyon in Tibet, a region of southwestern China, was formed over millions of years by the Yarlung Zangbo River. This canyon is the deepest in the world—at some points extending more than 5,300 meters (17,490 feet) from top to bottom. Yarlung Zangbo Canyon is also one of the world's longest canyons, at about 500 kilometers (310 miles).

Weathering and Erosion

Weathering and erosion also contribute to the formation of canyons. In winter, water seeps into cracks in the rock. This water freezes. As water freezes, it expands and turns into ice. Ice forces the cracks to become larger and larger, eroding bits of stone in the process. During brief, heavy rains, water rushes down the cracks, eroding even more rocks and stone. As more rocks crumble and fall, the canyon grows wider at the top than at the bottom.

When this process happens in soft rock, such as sandstone, it can lead to the development of slot canyons. Slot canyons are very narrow and deep. Sometimes, a slot canyon can be less than a meter (3 feet) wide, but hundreds of meters deep. Slot canyons can be dangerous. Their sides are usually very smooth and difficult to climb.

Some canyons with hard, underlying rock may develop cliffs and ledges after their softer, surface rock erodes. These ledges look like giant steps.

Sometimes, entire civilizations can develop on and around these canyon ledges. Native American nations, such as the Hopi and Sinagua, made cliff dwellings. Cliff dwellings were apartment-style shelters that housed hundreds of people. The shaded, elevated ledges in Walnut Canyon and Canyon de Chelly, in Arizona, provided protection from hostile neighbors and the burning desert sun.

Hard-rock canyons that are open at one end are called box canyons. The Hopi and Navajo people often used box canyons as natural corrals for sheep and cattle. They simply built a gate on the open side of the box canyon, and closed it when the animals were inside.

Limestone is a type of hard rock often found in canyons. Sometimes, limestone erodes and forms caves beneath the earth. As the ceilings of these caves collapse, canyons form. The Yorkshire Dales, an area in northern England, is a collection of river valleys and canyons created by limestone cave collapses.

Tectonic Uplift

Canyons are also formed by tectonic activity. As tectonic plates beneath the Earth's crust shift and collide, their movement can change the area's landscape. Sometimes, tectonic activity causes an area of the Earth's crust to rise higher than the surrounding land. This process is called tectonic uplift. Tectonic uplift can create plateaus and mountains. Rivers and glaciers that cut through these elevated areas of land create deep canyons.

The Grand Canyon, in the U.S. state of Arizona, is a product of tectonic uplift. The Grand Canyon, up to 447 kilometers (277 miles) long, 29 kilometers (18 miles) wide, and 1.8 kilometers (6,000 feet) deep, is the largest canyon in the United States. The Grand Canyon has been carved, over millions of years, as the Colorado River cuts through the Colorado Plateau. The Colorado Plateau is a large area that was elevated through tectonic uplift millions of years ago. Geologists debate the age of the canyon itself—it may be between 5 million and 70 million years old.

Canyons are like silent journals of an area's history over thousands or even millions of years. By studying the exposed layers of rock in a canyon wall, experts can learn about how the climate changed, what kind of organisms were alive at certain times, and perhaps even how the canyon may change in the future.

For example, geologists studying layers of rock in the Columbia River Gorge, in the U.S. states of Washington and Oregon, discovered that the oldest rocks there are at least 17 million years old. They also found out the rocks are dark-black basalt, made from hardened lava. From this, geologists determined that the rocks formed when volcanoes erupted and their lava spilled out onto the land. Over millions of years, the Columbia River and Ice Age glaciers cut through the area and exposed its volcanic beginnings.

Canyons are also important to paleontology, or the study of fossils. Fossils are often best preserved in dry, hot areas. Since canyons usually form under the same conditions, they are good places to examine fossils.

The layers of sediment revealed by a canyon can make it easier to date fossils. For example, a new area of dinosaur tracks was discovered in the U.S. state of Utah at Glen Canyon National

Recreation Area in 2009. These tracks reveal new information about a group of dinosaurs called ornithopods. Paleontologists analyzed the layers of rock surrounding the fossils to estimate how old they were. These new dinosaur tracks show that ornithopods were alive 20 million years earlier than scientists thought.

Geologists study canyons to determine how the landscape will change in the future. The erosion patterns and thickness of different layers can reveal the climate during different years. A series of very dry years will have very thin layers of rock, when little erosion took place. The overall pattern of erosion and layering reveals the rate of water flow, from both the river and rain, through a canyon.

Geologists estimate that the Grand Canyon, for example, is being eroded at a rate of 0.3 meters (1 foot) every 200 years. The Colorado Plateau, the geologic area where the Grand Canyon is located, is a very stable area. Geologists expect the Grand Canyon to continue to deepen as long as the Colorado River flows.

Submarine Canyons

Some of the deepest canyons lie beneath the ocean. These submarine canyons cut into continental shelves and continental slopes—the edges of continents that are underwater.

Some submarine canyons were carved by rivers that flowed during periods when the sea level was lower, and the continental shelves were exposed. The Hudson Canyon extends 750 kilometers (450 miles) into the Atlantic Ocean, from the mouth of the Hudson River, in the U.S. states of New York and New Jersey. At least part of the Hudson Canyon was the river bed during the last ice age, when sea levels were much lower.

Submarine canyons can also develop when powerful ocean currents sweep away sediments. Just as rivers erode land, these currents carve deep canyons in the ocean floor. Strong currents of the Atlantic Ocean prevent Whittard Canyon, about 400 kilometers (248 miles) south of the coast of Ireland, from filling with sediment. Scientists studying Whittard Canyon believe glacial water mixed with seawater to rush into the submarine canyon thousands of years ago.

The formation of some submarine canyons is still a mystery. Monterey Canyon is a deep submarine canyon off the coast of the U.S. state of California. It has been compared to the Grand Canyon because of its size. It is 152 kilometers (95 miles) long and 3.2 kilometers (2 miles) deep at its deepest point. Geologists still aren't certain how Monterey Canyon was formed. One theory is that the canyon was formed by an ancient outlet of the Sacramento or Colorado Rivers. Another theory is that it was formed by tectonic activity—an earthquake splitting apart the rock with enormous force. Scientists believe the canyon was formed 25 million to 30 million years ago.

The depth of submarine canyons makes them hard to explore. Scientists usually use remotely operated vehicles (ROVs) to conduct studies. Sometimes, they can use a submersible, a special kind of submarine. The Monterey Bay Aquarium Research Institute (MBARI) uses a vehicle called Ventana to explore Monterey Canyon. Through the Ventana and other research vehicles, MBARI scientists have discovered new species of organisms living in the canyon, from tiny sea anemones to giant squid.

The Grand Canyon, brought to you by the Colorado Plateau and the Colorado River.

Fluvial and Coastal Landforms

Coastal landforms are any of the relief features present along any coast, the result of a combination of processes, sediments, and the geology of the coast itself.

The coastal environment of the world is made up of a wide variety of landforms manifested in a spectrum of sizes and shapes ranging from gently sloping beaches to high cliffs, yet coastal landforms are best considered in two broad categories: erosional and depositional. In fact, the overall nature of any coast may be described in terms of one or the other of these categories. However, that each of the two major landform types may occur on any given reach of coast.

Factors and Forces in the Formation of Coastal Features

The landforms that develop and persist along the coast are the result of a combination of processes acting upon the sediments and rocks present in the coastal zone. The most prominent of these processes involves waves and the currents that they generate, along with tides. Other factors that significantly affect coastal morphology are climate and gravity.

Waves

The most obvious of all coastal processes is the continual motion of the waves moving toward the beach. Waves vary considerably in size over time at any given location and also vary markedly from place to place. Waves interact with the ocean bottom as they travel into shallow water; as a result, they cause sediment to become temporarily suspended and available for movement by coastal currents. The larger the wave, the deeper the water in which this process takes place and the larger the particle that can be moved. Even small waves that are only a few tens of centimeters high can pick up sand as they reach the shore. Larger waves can move cobbles and rock material as large as boulders.

Generally, small waves cause sediment—usually sand—to be transported toward the coast and to become deposited on the beach. Larger waves, typically during storms, are responsible for the removal of sediment from the coast and its conveyance out into relatively deep water.

Waves erode the bedrock along the coast largely by abrasion. The suspended sediment particles in waves, especially pebbles and larger rock debris, have much the same effect on a surface as sandpaper does. Waves have considerable force and so may break up bedrock simply by impact.

Longshore Currents

Waves usually approach the coast at some acute angle rather than exactly parallel to it. Because of this, the waves are bent (or refracted) as they enter shallow water, which in turn generates a current along the shore and parallel to it. Such a current is called a longshore current, and it extends from the shoreline out through the zone of breaking waves. The speed of the current is related to the size of the waves and to their angle of approach. Under rather quiescent conditions, longshore currents move only about 10–30 centimeters per second; however, under stormy conditions they may exceed one meter per second. The combination of waves and longshore current acts to transport large quantities of sediment along the shallow zone adjacent to the shoreline.

Because longshore currents are caused by the approaching and refracting waves, they may move in either direction along the coast, depending on the direction of wave approach. This direction of approach is a result of the wind direction, which is therefore the ultimate factor in determining the direction of longshore currents and the transport of sediment along the shoreline.

Although a longshore current can entrain sediment if it moves fast enough, waves typically cause sediment to be picked up from the bottom, and the longshore current transports it along the coast. In some locations there is quite a large volume of net sediment transport along the coast because of a dominance of one wind direction—and therefore wave direction—over another. This volume may be on the order of 100,000 cubic meters per year. Other locations may experience more of a balance in wave approach, which causes the longshore current and sediment transport in one direction to be nearly balanced by the same process in the other direction.

Rip Currents

Another type of coastal current caused by wave activity is the rip current (incorrectly called rip tide in popular usage). As waves move toward the beach, there is some net shoreward transport of water. This leads to a slight but important upward slope of the water level (setup), so that the absolute water level at the shoreline is a few centimeters higher than it is beyond the surf zone. This situation is an unstable one, and water moves seaward through the surf zone in an effort to relieve the instability of the sloping water. The seaward movement is typically confined to narrow pathways. In most cases, rip currents are regularly spaced and flow at speeds of up to several tens of centimeters per second. They can carry sediment and often are recognized by the plume of suspended sediment moving out through the surf zone. In some localities rip currents persist for months at the same site, whereas in others they are quite ephemeral.

Tides

The rise and fall of sea level caused by astronomical conditions is regular and predictable. There is a great range in the magnitude of this daily or semi-daily change in water level. Along some coasts the tidal range is less than 0.5 meter, whereas in the Bay of Fundy in southeastern Canada the maximum tidal range is just over 16 meters. A simple but useful classification of coasts is based solely on tidal range without regard to any other variable. Three categories have been established: micro-tidal (less than two meters), meso-tidal (two to four meters), and macro-tidal (more than four meters). Micro-tidal coasts constitute the largest percentage of the world's coasts, but the other two categories also are widespread.

The role of tides in molding coastal landforms is twofold: (1) tidal currents transport large quantities of sediment and may erode bedrock, and (2) the rise and fall of the tide distributes wave energy across a shore zone by changing the depth of water and the position of the shoreline.

Tidal currents transport sediment in the same way that longshore currents do. The speeds necessary to transport the sediment (typically sand) are generated only under certain conditions—usually in inlets, at the mouths of estuaries, or any other place where there is a constriction in the coast through which tidal exchange must take place. Tidal currents on the open coast, such as along a beach or rocky coast, are not swift enough to transport sediment. The speed of tidal currents in constricted areas, however, may exceed two meters per second, especially in inlets located on a barrier island complex. The speed of these tidal currents is dictated by the volume of water that must pass through the inlet during a given flood or ebb-tide cycle. This may be either six or 12 hours in duration, depending on whether the local situation is semidiurnal (12-hour cycle) or diurnal (24-hour cycle). The volume of water involved, called the tidal prism, is the product of the tidal range and the area of the coastal bay being served by the inlet. This means that though there may be a direct relationship between tidal range and tidal-current speed, it is also possible to have very swift tidal currents on a coast where the tidal range is low if the bay being served by the inlet is quite large. This is a very common situation along the coast of the Gulf of Mexico where the range is typically less than one meter but where there are many large coastal bays.

The rise and fall of the tide along the open coast has an indirect effect on sediment transport, even though currents capable of moving sediment are not present. As the tide comes in and then retreats along a beach or on a rocky coast, it causes the shoreline to move accordingly. This movement of the shoreline changes the zone where waves and longshore currents can do their work. Tidal range in combination with the topography of the coast is quite important in this situation. The greater the tidal range, the more effect this phenomenon has on the coast. The slope of a beach or other coastal landform also is important, however, because a steep cliff provides only a nominal change in the area over which waves and currents can do their work even in a macro-tidal environment. On the other hand, a broad, gently sloping beach or tidal flat may experience a change in the shoreline of as much as one kilometer during a tidal cycle in a macro-tidal setting. Examples of this situation occur in the Bay of Fundy and along the West German coast of the North Sea.

Other Factors and Processes

Climate is an extremely important factor in the development of coastal landforms. The elements of climate include rainfall, temperature, and wind.

Rainfall is important because it provides runoff in the form of streams and also is a factor in producing and transporting sediment to the coast. This fact gives rise to a marked contrast between the volume and type of sediment carried to the coast in a tropical environment and those in a desert environment.

Temperature is important for two quite different reasons. It is a factor in the physical weathering of sediments and rocks along the coast and in the adjacent drainage basins. This is particularly significant in cold regions where the freezing of water within cracks in rocks causes the rocks to fragment and thereby yield sediment. Some temperate and arctic regions have shore ice up to several months each year. Under these conditions there is no wave impact, and the coast becomes essentially static until the ice thaws or breaks up during severe storms. Such conditions prevail for three to four months along much of the coast of the Great Lakes in North America.

Wind is important primarily because of its relationship to waves. Coasts that experience prolonged and intense winds also experience high wave-energy conditions. Seasonal patterns in both wind direction and intensity can be translated directly into wave conditions. Wind also can be a key factor in directly forming coastal landforms, particularly coastal dunes. The persistence of onshore winds throughout much of the world's coast gives rise to sand dunes in all places where enough sediment is available and where there is a place for it to accumulate.

Gravity, too, plays a major role in coastal processes. Not only is it indirectly involved in processes associated with wind and waves but it also is directly involved through downslope movement of sediment and rock as well. This role is particularly evident along shoreline cliffs where waves attack the base of the cliffs and undercut the slope, resulting in the eventual collapse of rocks into the sea or their accumulation as debris at the base of the cliffs.

Landforms Of Erosional Coasts

There are two major types of coastal morphology: one is dominated by erosion and the other by deposition. They exhibit distinctly different landforms, though each type may contain some features of the other. In general, erosional coasts are those with little or no sediment, whereas depositional coasts are characterized by abundant sediment accumulation over the long term. Both temporal and geographic variations may occur in each of these coastal types.

Erosional coasts typically exhibit high relief and rugged topography. They tend to occur on the leading edge of lithospheric plates, the west coasts of both North and South America being excellent examples. Glacial activity also may give rise to erosional coasts, as in northern New England and in the Scandinavian countries. Typically, these coasts are dominated by exposed bedrock with steep slopes and high elevations adjacent to the shore. Although these coasts are erosional, the rate of shoreline retreat is slow due to the resistance of bedrock to erosion. The type of rock and its lithification are important factors in the rate of erosion.

Sea Cliffs

The most widespread landforms of erosional coasts are sea cliffs. These very steep to vertical bedrock cliffs range from only a few meters high to hundreds of meters above sea level. Their vertical nature is the result of wave-induced erosion near sea level and the subsequent collapse of rocks at higher elevation. Cliffs that extend to the shoreline commonly have a notch cut into them where waves have battered the bedrock surface.

At many coastal locations there is a thin, narrow veneer of sediment forming a beach along the base of sea cliffs. This sediment may consist of sand, but it is more commonly composed of coarse material—cobbles or boulders. Beaches of this kind usually accumulate during relatively low wave-energy conditions and are removed during the stormy season when waves are larger. The coasts of California and Oregon contain many places where this situation prevails. The presence of even a narrow beach along a rocky coast provides the cliffs protection against direct wave attack and slows the rate of erosion.

Wave-cut Platforms

At the base of most cliffs along a rocky coast one finds a flat surface at about the mid-tide elevation. This is a bench like feature called a wave-cut platform, or wave-cut bench. Such surfaces may measure from a few meters to hundreds of meters wide and extend to the base of the adjacent cliff. They are formed by wave action on the bedrock along the coast. The formation process can take a long time, depending on the type of rock present. The existence of extensive wave-cut platforms thus implies that sea level did not fluctuate during the periods of formation. Multiple platforms of this type along a given reach of coast indicate various positions of sea level.

Sea Stacks

Erosion along rocky coasts occurs at various rates and is dependent both on the rock type and on the wave energy at a particular site. As a result of the above-mentioned conditions, wave-cut platforms may be incomplete, with erosional remnants on the horizontal wave-cut surface. These remnants are called sea stacks, and they provide a spectacular type of coastal landform. Some are many meters high and form isolated pinnacles on the otherwise smooth wave-cut surface. Because erosion is a continual process, these features are not permanent and will eventually be eroded, leaving no trace of their existence.

Sea Arches

Another type of erosional landform is the sea arch, which forms as the result of different rates of erosion typically due to the varied resistance of bedrock. These archways may have an arcuate or rectangular shape, with the opening extending below water level. The height of an arch can be up to tens of meters above sea level.

It is common for sea arches to form when a rocky coast undergoes erosion and a wave-cut platform develops. Continued erosion can result in the collapse of an arch, leaving an isolated sea stack on the platform. Still further erosion removes the stack, and eventually only the wave-cut platform remains adjacent to the eroding coastal cliff.

Landforms Of Depositional Coasts

Coasts adjacent to the trailing edge of lithospheric plates tend to have widespread coastal plains and low relief. The Atlantic and Gulf coasts of the United States are representative. Such coasts may have numerous estuaries and lagoons with barrier islands or may develop river deltas. They are characterized by an accumulation of a wide range of sediment types and by many varied coastal environments. The sediment is dominated by mud and sand; however, some gravel may be present, especially in the form of shell material.

Depositional coasts may experience erosion at certain times and places due to such factors as storms, depletion of sediment supply, and rising sea level. The latter is a continuing problem as the mean annual temperature of the Earth rises and the ice caps melt. Nevertheless, the overall, long-range tendency along these coasts is that of sediment deposition.

Waves, wave-generated currents, and tides significantly influence the development of depositional landforms. In general, waves exert energy that is distributed along the coast essentially parallel to it. This is accomplished by the waves themselves as they strike the shore and also by the longshore currents that move along it. In contrast, tides tend to exert their influence perpendicular to the coast as they flood and ebb. The result is that the landforms that develop along some coasts are due primarily to wave processes while along other coasts they may be due mainly to tidal processes. Some coasts are the result of near equal balance between tide and wave processes. As a consequence, investigators speak of wave-dominated coasts, tide-dominated coasts, and mixed coasts.

A wave-dominated coast is one that is characterized by well-developed sand beaches typically formed on long barrier islands with a few widely spaced tidal inlets. The barrier islands tend to be narrow and rather low in elevation. Longshore transport is extensive, and the inlets are often small and unstable. Jetties are commonly placed along the inlet mouths to stabilize them and keep them open for navigation. The Texas and North Carolina coasts of the United States are excellent examples of this coastal type.

Tide-dominated coasts are not as widespread as those dominated by waves. They tend to develop where tidal range is high or where wave energy is low. The result is a coastal morphology that is dominated by funnel-shaped embayments and long sediment bodies oriented essentially perpendicular to the overall coastal trend. Tidal flats, salt marshes, and tidal creeks are extensive. The West German coast of the North Sea is a good example of such a coast.

Mixed coasts are those where both tidal and wave processes exert considerable influence. These coasts characteristically have short stubby barrier islands and numerous tidal inlets. The barriers commonly are wide at one end and narrow at the other. Inlets are fairly stable and have large sediment bodies on both their landward and seaward sides. The Georgia and South Carolina coasts of the United States typify a mixed coast.

General Coastal Morphology

Depositional coasts can be described in terms of three primary large-scale types: (1) deltas, (2) barrier island/estuarine systems, and (3) strand-plain coasts. The latter two have numerous features in common.

Barrier Island/Estuarine Systems

Many depositional coasts display a complex of environments and landforms that typically occur together. Irregular coasts have numerous embayments, many of which are fed by streams. Such embayments are called estuaries, and they receive much sediment due to runoff from an adjacent coastal plain. Seaward of the estuaries are elongate barrier islands that generally parallel the shore. Consisting mostly of sand, they are formed primarily by waves and longshore currents. These barrier islands are typically separated from the mainland and may have lagoons, which are long, narrow, coastal bodies of water situated between the barrier and the mainland.

Most barrier islands contain a well-developed beach, coastal dunes, and various environments on their landward side, including tidal flats, marshes, or wash over fans. Such coastal barriers are typically interrupted by tidal inlets, which provide circulation between the various coastal bays and the open marine environment. These inlets also are important pathways for organisms that migrate between coastal and open marine areas as well as for pleasure and commercial boat traffic.

Strand-plain Coasts

Some wave-dominated coasts do not contain estuaries and have no barrier island system. These coasts, however, do have beaches and dunes, and may even have coastal marshes. The term strand plain has been applied to coasts of this sort. Examples include parts of western Louisiana and eastern Texas. In most respects, they are similar in morphology to barrier islands but lack inlets.

Beaches and Coastal Dunes

There are several specific landforms representative of coastal environments that are common to each of the three major categories described above. Especially prominent among these are beaches and dunes. They are the primary landforms on barrier islands, strand-plain coasts, and many deltas, particularly the wave-dominated variety.

Beaches

A consideration of the beach must also include the seaward adjacent near shore environment because the two are intimately related. The near shore environment extends from the outer limit of the longshore bars that are usually present to the low-tide line. In areas where longshore bars are absent, it can be regarded as coincident with the surf zone. The beach extends from the low-tide line to the distinct change in slope and/or material landward of the unvegetated and active zone of sediment accumulation. It may consist of sand, gravel, or even mud, though sand is the most common beach material.

The beach profile typically can be divided into two distinct parts: (1) the seaward and relatively steep sloping foreshore, which is essentially the intertidal beach, and (2) the landward, nearly horizontal backshore. Beach profiles take on two different appearances, depending on conditions at any given time. During quiescent wave conditions, the beach is said to be accretional, and both the foreshore and backshore are present. During storm conditions, however, the beach experiences erosion, and the result is typically a profile that shows only the seaward sloping foreshore. Because the beach tends to repair itself during nonstorm periods, a cyclic pattern of profile shapes is common.

The near shore zone is where waves steepen and break, and then re-form in their passage to the beach, where they break for the last time and surge up the foreshore. Much sediment is transported in this zone, both along the shore and perpendicular to it. During storms the waves tend to be steep, and erosion of the beach occurs with sediment transported offshore. The intervening calmer conditions permit sediment to be transported landward and rebuild the beach. Because wave conditions may change daily, the nature of the profile and the sediment on the foreshore portion of the beach may also change daily. This is the zone of continual change on the beach.

The backshore of the beach is not subjected to wave activity except during storm conditions. It is actually in the supra-tidal zone—i.e., the zone above high tide where inundation by water is caused not by regular astronomical tides but rather by storm-generated tides. During nonstorm conditions the back-beach is relatively inactive except for wind action, which may move sediment. In most cases, there is an onshore component to the wind, and sediment is carried from the back-beach landward, typically forming dunes. Any obstruction on the back-beach, such as vegetation, pieces of driftwood, fences, or even trash discarded by people, results in wind-blown sand accumulation.

There are variations in beach forms along the shore as well as in those perpendicular to the shore. Most common is the rhythmic topography that is seen along the foreshore. A close look at the shoreline along most beaches will show that it is not straight or gently curved but rather that it displays a regularly undulating surface much like a low-amplitude sine curve. This is seen both on the plan view of the shoreline and the topography of the foreshore. The spacing is regular along a given reach of coast, but it may vary from place to place or from time to time at a given place. At some locations, concentrations of gravel or shells may develop, forming beach cusps (more or less triangular deposits that point seaward) during some wave conditions.

Although there is a common trend to the beach profile, some variation exists both because of energy conditions and because of the material making up the beach. Generally speaking, a beach that is accumulating sediment and experiencing low-energy conditions tends to have a steep foreshore, whereas the same beach would have a relatively gentle foreshore during storm conditions when erosion is prevalent. The grain size of beach sediment also is an important factor in the slope of the foreshore. In general, the coarser the constituent grains, the steeper the foreshore. Examples include the gravel beaches of New England, as contrasted to the gently sloping sand beaches of the Texas coast.

Coastal Dunes

Immediately landward of the beach are commonly found large, linear accumulations of sand known as dunes. (For coverage of dunes in arid and semiarid regions, see sand dune.) They form as the wind carries sediment from the beach in a landward direction and deposits it wherever an obstruction hinders further transport. Sediment supply is the key limiting factor in dune development and is the primary reason why some coastal dunes, such as those on the west Florida peninsula, are quite small, whereas others in such areas as the Texas coast and the Florida panhandle have large dunes.

Small wind-shadow dunes, or coppice mounds, actually may form on the backshore of the beach. If sediment continues to be supplied and beach erosion does not destroy them, these small sand accumulations will become foredunes, the seaward-most line of coastal dunes. It is in this fashion that a coast progrades, or grows seaward. Many barrier-island or strand-plain coasts exhibit numerous, essentially parallel dune ridges testifying to this type of growth.

The sediment in dunes tends to be fine to medium sand that is quite well sorted. Shell debris or other material is uncommon unless it is the same size or mass as the dune sand. There are various types of vegetation that grow on the dune surface and stabilize it. These grasses and vines often can be seen on the backshore portion of beaches as well. Dunes lacking vegetation are usually active and exhibit various signs of sand mobility. Most widespread are the nearly ubiquitous ripples that cover the dune surface. Large lobes of sand or even an entire dune may also move as wind blows across the dune. This activity results in cross stratification of the dune in large sweeping patterns of wedge-shaped packages of sand.

River Delta

A river delta is a low-lying plain or landform that occurs at the mouth of a river near where it flows into an ocean or another larger body of water. Deltas' greatest importance to human activities, fish and wildlife lay in their characteristic highly fertile soil and dense, diverse vegetation.

In order to fully appreciate the role deltas play in our larger ecosystem, it is first important to understand rivers. Rivers are defined as bodies of fresh water generally flowing from high elevations toward an ocean, a lake or another river; sometimes, even back into the ground.

Most rivers begin at high elevations where snow, rain, and other precipitation run downhill into creeks and small streams. These small waterways flow ever farther downhill, eventually meeting to form rivers.

Rivers flow toward oceans or other larger bodies of water, oftentimes combining with other rivers. Deltas exist as the lowest part of these rivers. It is in these deltas where a river's flow slows and spreads out to create sediment-rich dry areas and bio diverse wetlands.

Formation of River Deltas

The formation of a river delta is a slow process. As rivers flow toward their outlets from higher elevations, they deposit mud, silt, sand, and gravel particles at the mouths where rivers and larger, more sedentary bodies of water meet.

Over time these particles (called sediment or alluvium) build up at the mouth, extending into the ocean or lake. As these areas continue to grow the water becomes shallower and eventually, landforms begin to rise above the surface of the water, typically elevating to just above sea level.

As rivers drop enough sediment to create these landforms or areas of raised elevation, the remaining flowing water with the most power sometimes cuts across the land, forming different branches called distributaries.

Once formed, deltas are typically made up of three parts: the upper delta plain, the lower delta plain, and the subaqueous delta.

The upper delta plain makes up the area nearest to land. It is usually the area with the least water and highest elevation.

The lower delta plain is the middle of the delta. It is a transition zone between the dry upper delta and the wet subaqueous delta.

The subaqueous delta is the portion of the delta closest to the sea or body of water into which the river flows. This area is usually past the shoreline and it is below water level.

Types of River Deltas

Despite the generally universal processes by which river deltas are formed and organized, it is important to note that the world's deltas vary dramatically in structure, composition, and size due to factors such as origin, climate, geology, and tidal processes. These external factors contribute to an impressive diversity of deltas around the world. A delta's characteristics are classified based upon the specific factors contributing to its river's deposition of sediment -- typically the river itself, waves or tides.

The main types of deltas are wave-dominated deltas, tide-dominated deltas, Gilbert deltas, inland deltas, and estuaries.

As its name would imply, a wave-dominated delta such as the Mississippi River Delta is created by wave erosion controlling where and how much river sediment remains in a delta once it has been dropped. These deltas are usually shaped like the Greek symbol, delta (Δ).

Tide-dominated deltas such as the Ganges River Delta are formed by tides. Such deltas are characterized by a dendritic structure (branched, like a tree) due to newly-formed distributaries during times of high water.

Gilbert deltas are steeper and formed by deposition of coarse material. While it is possible for them to form in ocean areas, their formations are more commonly seen in mountainous areas where mountain rivers deposit sediment into lakes.

Inland deltas are deltas formed in inland areas or valleys where rivers may divide into many branches and rejoin farther downstream. Inland deltas, also called inverted river deltas, normally form on former lake beds.

Finally, when a river is located near coasts characterized by large tidal variations, they do not always form a traditional delta. Tidal variation often results in estuaries or a river that meets the sea, such as Saint Lawrence River in Ontario, Quebec, and New York.

Humans and River Deltas

River deltas have been important to humans for thousands of years because of their extremely fertile soils. Major ancient civilizations grew along deltas such as those of the Nile and the Tigris-Euphrates rivers, with the inhabitants of these civilizations learning how to live with their natural flooding cycles.

Many people believe that the ancient Greek historian Herodotus first coined the term delta nearly 2,500 years ago as many deltas are shaped like the Greek delta (Δ) symbol.

Deltas remain important to humans even today as, among many other things, a source of sand and gravel. Used in highway, building and infrastructure construction, these highly valuable materials quite literally build our world.

Delta land is also important in agricultural use. Witness the Sacramento-San Joaquin Delta in California. One of the most agriculturally diverse and productive areas in the state, the region successfully supports numerous crops from kiwi to alfalfa to tangerines.

Biodiversity and Importance of River Deltas

In addition to these human uses, river deltas boast some of the most bio diverse systems on the planet. As such, it is essential that these unique and beautiful havens of biodiversity remain as healthy habitat for the many species of plants, animals, insects, and fish -- some rare, threatened or endangered -- which call them home.

In addition to their biodiversity, deltas and wetlands provide a buffer for hurricanes, as open land often stands to weaken the impact of storms as they travel toward larger, more populated areas. The Mississippi River Delta, for example, buffers the impact of potentially strong hurricanes in the Gulf of Mexico.

Peninsula

A peninsula is a piece of land that is almost entirely surrounded by water but is connected to the mainland on one side.

Peninsulas can be very small, sometimes only large enough for a single lighthouse, for instance. Lighthouses often sit on peninsulas near rocky coastlines to warn sailors that they are getting close to land.

Peninsulas can also be very large. Most of the U.S. state of Florida is a peninsula that separates the Gulf of Mexico and the Atlantic Ocean.

Peninsulas are found on every continent. In North America, the narrow peninsula of Baja California, in Mexico, separates the Pacific Ocean and the Sea of Cortez, also called the Gulf of California. In Europe, the nations of Portugal and Spain make up the Iberian Peninsula. The so-called Horn of Africa, which juts into the Arabian Sea on central Africa's east coast, is a huge peninsula. The nations of North Korea and South Korea make up the Korean Peninsula in eastern Asia. In Australia, the Cape York Peninsula is only 160 kilometers (99 miles) from the island of New Guinea. The Antarctic Peninsula seems to point to the tip of South America, several hundred kilometers (miles) away.

This peninsula juts into the Black Sea.

Meander

A meander is one of a series of regular sinuous curves, bends, loops, turns, or windings in the channel of a river, stream, or other watercourse. It is produced by a stream or river swinging from side to side as it flows across its floodplain or shifts its channel within a valley. A meander is produced by a stream or river as it erodes the sediments comprising an outer, concave bank (cut bank) and deposits this and other sediment downstream on an inner, convex bank which is typically a point bar. The result of sediments being eroded from the outside concave bank and their deposition on an inside convex bank is the formation of a sinuous course as a channel migrates back and forth across the down-valley axis of a floodplain. The zone within which a meandering stream shifts its channel across either its floodplain or valley floor from time to time is known as a meander belt. It typically ranges from 15 to 18 times the width of the channel. Over time, meanders migrate downstream, sometimes in such a short time as to create civil engineering problems for local municipalities attempting to maintain stable roads and bridges.

The degree of meandering of the channel of a river, stream, or other watercourse is measured by its sinuosity. The sinuosity of a watercourse is the ratio of the length of the channel to the straight line down-valley distance. Streams or rivers with a single channel and sinuosities of 1.5 or more are defined as meandering streams or rivers.

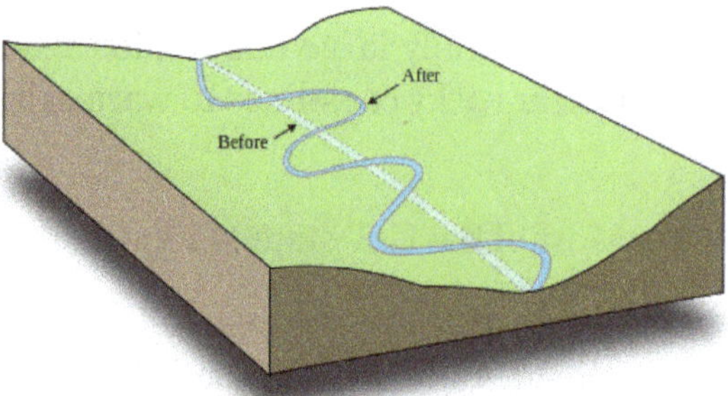

A streambed following a tilted valley. The maximum gradient is along the down-valley axis represented by a hypothetical straight channel. Meanders develop, which lengthen the course of the stream, decreasing the gradient.

Meanders of the *Rio Cauto* at Guamo Embarcadero, Cuba

Meanders of the Vydrica river in Bratislava forest park, Slovakia.

Governing Physics

When a fluid is introduced to an initially straight channel which then bends, the sidewalls induce a pressure gradient that causes the fluid to alter course and follow the bend. From here, two opposing processes occur: (1) irrotational flow and (2) secondary flow. For a river to 'meander', secondary flow must dominate.

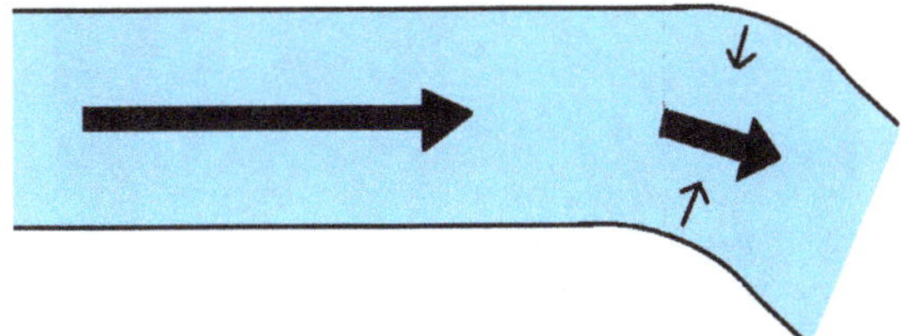

Straight channel culminating in a single bend

Irrotational flow: From Bernoulli's equations, high pressure results in low velocity. Therefore, *in the absence of secondary flow* we would expect low fluid velocity at the outside bend and high fluid velocity at the inside bend. This classic fluid mechanics result is *irrotational vortex flow*. In the context of meandering rivers, its effects are dominated by those of secondary flow.

Secondary flow: A force balance exists between pressure forces pointing to the inside bend of the river and centrifugal forces pointing to the outside bend of the river. In the context of meandering rivers, a boundary layer exists within the thin layer of fluid that interacts with the river bed. Inside that layer and following standard boundary-layer theory, the velocity of the fluid is effectively zero. Centrifugal force, which depends on velocity, is also therefore effectively zero. Pressure force, however, remains unaffected by the boundary layer. Therefore, within the boundary layer, pressure force dominates and fluid moves along the bottom of the river from the outside bend to the inside bend. This initiates helicoidal flow: Along the river bed, fluid roughly follows the curve of the channel but is also forced toward the inside bend; away from the river bed, fluid also roughly follows the curve of the channel but is forced, to some extent, from the inside to the outside bend. Ultimately, the downstream velocity of the fluid is convectively transported to the outside bend, resulting in higher velocities at the outside bend. This secondary flow effect dominates over that of irrotational flow: In real meandering rivers, we observe higher downstream fluid velocities at the outside bends.

The higher (lower) velocities at the outside (inside) bend result in higher (lower) shear stresses and therefore results in erosion (deposition). Thus meander bends erode at the outside bend, causing the river to becoming increasingly sinuous (until cutoff events occur). Deposition at the inside bend occur such that for most natural meandering rivers, the river width remains nearly constant, even as the river evolves.

Meander Geometry

Uvac canyon meander, Serbia

Meanders on the River Clyde, Scotland

The technical description of a meandering watercourse is termed meander geometry or meander planform geometry. It is characterized as an irregular waveform. Ideal waveforms, such as a sine wave, are one line thick, but in the case of a stream the width must be taken into consideration. The bank full width is the distance across the bed at an average cross-section at the full-stream level, typically estimated by the line of lowest vegetation.

As a waveform the meandering stream follows the down-valley axis, a straight line fitted to the curve such that the sum of all the amplitudes measured from it is zero. This axis represents the overall direction of the stream.

At any cross-section the flow is following the sinuous axis, the centerline of the bed. Two consecutive crossing points of sinuous and down-valley axes define a meander loop. The meander is two consecutive loops pointing in opposite transverse directions. The distance of one meander along the down-valley axis is the meander length or wavelength. The maximum distance from the down-valley axis to the sinuous axis of a loop is the meander width or amplitude. The course at that point is the apex.

In contrast to sine waves, the loops of a meandering stream are more nearly circular. The curvature varies from a maximum at the apex to zero at a crossing point (straight line), also called an inflection, because the curvature changes direction in that vicinity. The radius of the loop is the straight line perpendicular to the down-valley axis intersecting the sinuous axis at the apex. As the loop is not ideal, additional information is needed to characterize it. The orientation angle is the angle between sinuous axis and down-valley axis at any point on the sinuous axis.

Concave bank and convex bank, Great Ouse Relief Channel, England.

A loop at the apex has an outer or concave bank and an inner or convex bank. The meander belt is defined by an average meander width measured from outer bank to outer bank instead of from centerline to centerline. If there is a flood plain, it extends beyond the meander belt. The meander is then said to be free—it can be found anywhere in the flood plain. If there is no flood plain, the meanders are fixed.

Various mathematical formulae relate the variables of the meander geometry. As it turns out some numerical parameters can be established, which appear in the formulae. The waveform depends ultimately on the characteristics of the flow but the parameters are independent of it and apparently are caused by geologic factors. In general the meander length is 10–14 times, with an average 11 times, the full bank channel width and 3 to 5 times, with an average of 4.7 times, the radius of curvature at the apex. This radius is 2–3 times the channel width.

Meander of the River Cuckmere in Southern England

A meander has a depth pattern as well. The cross-overs are marked by riffles, or shallow beds, while at the apices are pools. In a pool direction of flow is downward, scouring the bed material. The major volume, however, flows more slowly on the inside of the bend where, due to decreased velocity, it deposits sediment.

The line of maximum depth, or channel, is the thalweg or thalweg line. It is typically designated the borderline when rivers are used as political borders. The thalweg hugs the outer banks and returns to center over the riffles. The meander arc length is the distance along the thalweg over one meander. The river length is the length along the centerline.

Formation

Meander formation is a result of natural factors and processes. The waveform configuration of a stream is constantly changing. Fluid flows around a bend in a vortex. Once a channel begins to follow a sinusoidal path, the amplitude and concavity of the loops increase dramatically due to the effect of helical flow sweeping dense eroded material towards the inside of the bend, and leaving the outside of the bend unprotected and therefore vulnerable to accelerated erosion, forming a positive feedback loop. In the words of Elizabeth A. Wood:

> 'this process of making meanders seems to be a self-intensifying process in which greater curvature results in more erosion of the bank, which results in greater curvature'

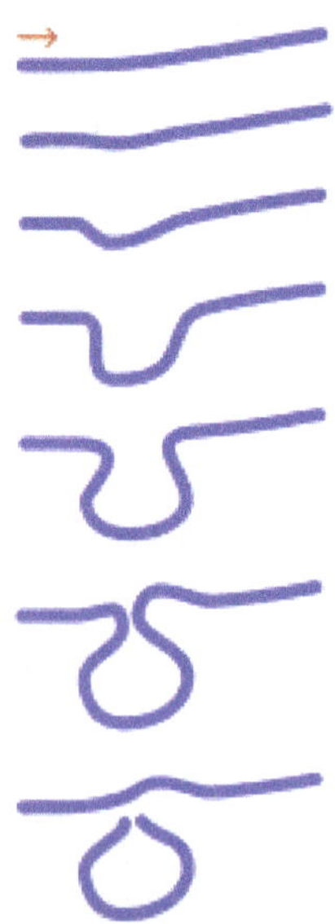

Life history of a meander

The crosscurrent along the floor of the channel is part of the secondary flow and sweeps dense eroded material towards the inside of the bend. The cross-current then rises to the surface near the inside and flows towards the outside, forming the helical flow. The greater the curvature of the bend, and the faster the flow, the stronger is the crosscurrent and the sweeping.

Due to the conservation of angular momentum the speed on the inside of the bend is faster than on the outside.

Since the flow velocity is diminished, so is the centrifugal pressure. However, the pressure of the super-elevated column prevails, developing an unbalanced gradient that moves water back across the bottom from the outside to the inside. The flow is supplied by a counter-flow across the surface from the inside to the outside. This entire situation is very similar to the Tea leaf paradox. This secondary flow carries sediment from the outside of the bend to the inside making the river more meandering.

As to why streams of any size become sinuous in the first place, there are a number of theories, not necessarily mutually exclusive.

Stochastic Theory

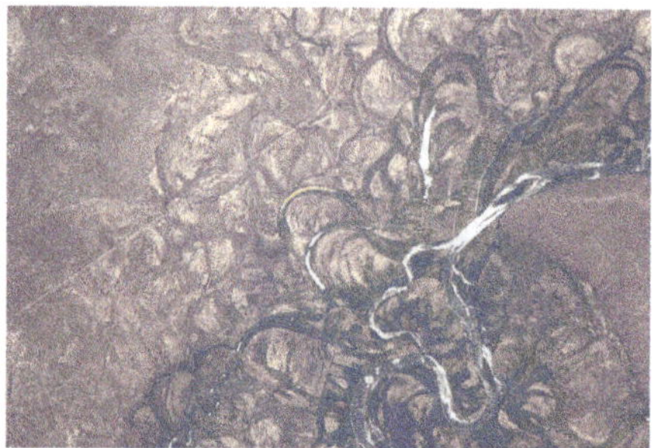

Meander scars, oxbow lakes and abandoned meanders in the broad flood plain of the Rio Negro, Argentina astronaut photo from ISS.

The stochastic theory can take many forms but one of the most general statements is that of Scheidegger: 'The meander train is assumed to be the result of the stochastic fluctuations of the direction of flow due to the random presence of direction-changing obstacles in the river path.' Given a flat, smooth, tilted artificial surface, rainfall runs off it in sheets, but even in that case adhesion of water to the surface and cohesion of drops produce rivulets at random. Natural surfaces are rough and erodible to different degrees. The result of all the physical factors acting at random is channels that are not straight, which then progressively become sinuous. Even channels that appear straight have a sinuous thalweg that leads eventually to a sinuous channel.

Equilibrium Theory

In the equilibrium theory, meanders decrease the stream gradient until an equilibrium between the erodibility of the terrain and the transport capacity of the stream is reached. A mass of water descending must give up potential energy, which, given the same velocity at the end of the drop as at the beginning, is removed by interaction with the material of the streambed. The shortest distance; that is, a straight channel, results in the highest energy per unit of length, disrupting the banks more, creating more sediment and aggrading the stream. The presence of meanders allows the stream to adjust the length to an equilibrium energy per unit length in which the stream carries away all the sediment that it produces.

Geomorphic and Morphotectonic Theory

Geomorphic refers to the surface structure of the terrain. Morphotectonic means having to do with the deeper, or tectonic (plate) structure of the rock. The features included under these categories are not random and guide streams into non-random paths. They are predictable obstacles that instigate meander formation by deflecting the stream. For example, the stream might be guided into a fault line (morphotectonic).

Associated Landforms

Cut Bank

A cut bank is an often-vertical bank or cliff that forms where the outside, concave bank of a meander cuts into the floodplain or valley wall of a river or stream. A cutbank is also known either as a river-cut cliff, river cliff, or a bluff and spelled as cutbank. Erosion that forms a cut bank occurs at the outside bank of a meander because helicoidal flow of water keeps the bank washed clean of loose sand, silt, and sediment and subjects it to constant erosion. As a result, the meander erodes and migrates in the direction of the outside bend, forming the cut bank.

As the cut bank is undermined by erosion, it commonly collapses as slumps into the river channel. The slumped sediment, having been broken up by slumping, is readily eroded and carried toward the middle of the channel. The sediment eroded from a cut bank tends to be deposited on the point bar of the next downstream meander, and not on the point bar opposite it. This can be seen in areas where trees grow on the banks of rivers; on the inside of meanders, trees, such as willows, are often far from the bank, whilst on the outside of the bend, the tree roots are often exposed and undercut, eventually leading the trees to fall into the river.

Meander Cutoff

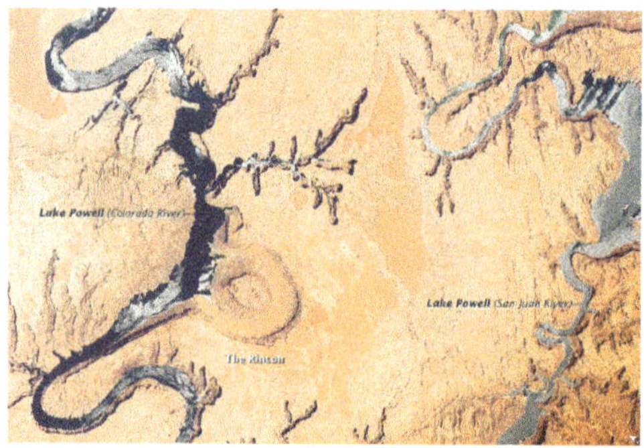

The Rincon on Lake Powell in southern Utah. It is an incised cutoff (abandoned) meander.

A meander cutoff, also known as either a cutoff meander or abandoned meander, is a meander that has been abandoned by its stream after the formation of a neck cutoff. A lake that occupies an cutoff meander is known as an *oxbow lake*. Cutoff meanders that have cut downward into the underlying bedrock are known in general as incised cutoff meanders. As in the case of the Anderson Bottom Rincon, incised meanders that have either steep-sided, often vertical walls, are often, but not always, known as rincons in the southwest United States. *Rincon* in English is a nontechnical word in the southwest United States for either a small-secluded valley, an alcove or angular recess in a cliff, or a bend in a river.

Incised Meanders

Glen Canyon, US

The meanders of a stream or river that has cut its bed down into the bedrock are known as either incised, intrenched, entrenched, inclosed or ingrown meanders. Some Earth scientists recognize and use a finer subdivision of incised meanders. Thornbury argues that *incised* or *inclosed meanders* are synonyms that are appropriate to describe any meander incised downward into bedrock and defines *enclosed* or *entrenched meanders* as a subtype of incised meanders (inclosed meanders) characterized by a symmetrical valley sides. He argues that the symmetrical valley sides are the direct result of rapid down cutting of a watercourse into bedrock. In addition, as proposed by Rich, Thornbury argues that incised valleys with a pronounced asymmetry of cross section, which he called *ingrown meanders*, are the result of the lateral migration and incision of a meander during a period of slower channel down cutting. Regardless, the formation of both entrenched meanders and ingrown meanders is thought to require that base level falls as a result of either relative change in mean sea level, isostatic or tectonic uplift, the breach of an ice or landslide dam, or regional tilting. Classic examples of incised meanders are associated with rivers in the Colorado Plateau, the Kentucky River Palisades in central Kentucky, and streams in the Ozark Plateau.

Goosenecks of the San Juan River, SE Utah.

It was initially either argued or presumed that an incised meander is characteristic of an antecedent stream or river that had incised its channel into underlying strata. An antecedent stream or river is one that maintains its original course and pattern during incision despite the changes in underlying rock topography and rock types. However, later geologists argue that the shape of an incised meander is not always, if ever, "inherited," e.g., strictly from an antecedent meandering stream where it meander pattern could freely develop on a level floodplain. Instead, they argue that as fluvial incision of bedrock proceeds, the stream course is significantly modified by variations in rock type and fractures, faults, and other geological structures into either *lithologically conditioned meanders* or *structurally controlled meanders*.

Oxbow Lakes

The oxbow lake, which is the most common type of fluvial lake, is a crescent-shaped lake that derives its name from its distinctive curved shape. Oxbow lakes are also known as cutoff lakes. Such lakes form regularly in undisturbed floodplains as a result of the normal process of fluvial meandering. Either a river or stream forms a sinuous channel as the outer side of its bends are eroded away and sediments accumulate on the inner side, which forms a meandering horseshoe-shaped bend. Eventually as the result of its meandering, the fluvial channel cuts through the narrow neck of the meander and forms a cutoff meander. The final break-through of the neck, which is called a neck cutoff, often occurs during a major flood because that is when the watercourse is out of its banks and can flow directly across the neck and erode it with the full force of the flood.

After a cutoff meander is formed, river water flows into its end from the river builds small delta-like feature into either end of it during floods. These delta-like features block either end of the cutoff meander to form a stagnant oxbow lake that is separated from the flow of the fluvial channel and independent of the river. During floods, the flood waters deposit fine-grained sediment into the oxbow lake. As a result, oxbow lakes tend to become filled in with fine-grained, organic-rich sediments over time.

Point Bar

A point bar, which is also known as a meander bar, is a fluvial bar that is formed by the slow, often episodic, addition of individual accretions of noncohesive sediment on the inside bank of a meander by the accompanying migration of the channel toward its outer bank. This process is called lateral accretion. Lateral accretion occurs mostly during high water or floods when the point bar is submerged. Typically, the sediment consists of either sand, gravel, or a combination of both. The sediment comprising some point bars might grade downstream into silty sediments. Because of the decreasing velocity and strength of current from the thalweg of the channel to the upper surface of point bar when the sediment is deposited the vertical sequence of sediments comprising a point bar becomes finer upward within an individual point bar. For example, it is typical for point bars to fine upward from gravel at the base to fine sands at the top. The source of the sediment is typically upstream cut banks from which sand, rocks and debris has been eroded, swept, and rolled across the bed of the river and downstream to the inside bank of a river bend. On the inside bend, this sediment and debris is eventually deposited on the slip-off slope of a point bar.

Scroll-bars

Scroll-bars are a result of continuous lateral migration of a meander loop that creates an asymmetrical ridge and swale topography on the inside of the bends. The topography is generally parallel to the meander, and is related to migrating bar forms and back bar chutes, which carve sediment from the outside of the curve and deposit sediment in the slower flowing water on the inside of the loop, in a process called lateral accretion. Scroll-bar sediments are characterized by cross bedding and a pattern of fining upward. These characteristics are a result of the dynamic river system, where larger grains are transported during high energy flood events and then gradually die down, depositing smaller material with time. Deposits for meandering rivers are generally homogeneous and laterally extensive unlike the more heterogeneous braided river deposits. There are two distinct patterns of scroll-bar depositions; the eddy accretion scroll bar pattern and the point-bar scroll pattern. When looking down the river valley they can be distinguished because the point-bar scroll patterns are convex and the eddy accretion scroll bar patterns are concave.

Scroll bars often look lighter at the tops of the ridges and darker in the swales. This is because the tops can be shaped by wind, either adding fine grains or by keeping the area unvegetated, while the darkness in the swales can be attributed to silts and clays washing in during high water periods. This added sediment in addition to water that catches in the swales is in turn is a favorable environment for vegetation that will also accumulate in the swales.

Slip-off Slope

Depending upon whether a meander is part of an entrenched river or part of a freely meandering river within a floodplain, the term slip-off slope can refer to two different fluvial landforms that comprise the inner, convex, bank of a meander loop. In case of a freely meandering river on a floodplain, a *slip-off slope* is the inside, gently sloping bank of a meander on which sediments episodically accumulate to form a point bar as a river meanders. This type of slip-off slope is located opposite the cutbank. This term can also be applied to the inside, sloping bank of a meandering tidal channel.

In case of an entrenched river, a *slip-off slope* is a gently sloping bedrock surface that rises from the inside, concave bank of an asymmetrically entrenched river. This type of slip-off slope is often covered by a thin, discontinuous layer of alluvium. It is produced by the gradual outward migration of the meander as a river cuts downward into bedrock. A terrace on the slip-off slope of a meander spur, known as slip-off slope terrace, can formed by a brief halt during the irregular incision by an actively meandering river.

Derived Quantities

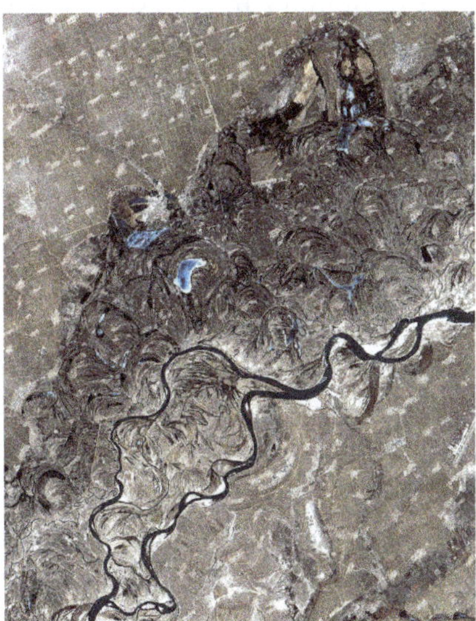

Meanders, scroll-bars and oxbow lakes in the Songhua River

The meander ratio or sinuosity index is a means of quantifying how much a river or stream meanders (how much its course deviates from the shortest possible path). It is calculated as the length of the stream divided by the length of the valley. A perfectly straight river would have a meander ratio of 1 (it would be the same length as its valley), while the higher this ratio is above 1, the more the river meanders.

Sinuosity indices are calculated from the map or from an aerial photograph measured over a distance called the reach, which should be at least 20 times the average full bank channel width. The length of the stream is measured by channel, or thalweg, length over the reach, while the bottom value of the ratio is the down valley length or air distance of the stream between two points on it defining the reach.

The sinuosity index plays a part in mathematical descriptions of streams. The index may require elaboration, because the valley may meander as well—i.e., the down valley length is not identical to the reach. In that case the valley index is the meander ratio of the valley while the channel index is the meander ratio of the channel. The channel sinuosity index is the channel length divided by the valley length and the standard sinuosity index is the channel index divided by the valley index. Distinctions may become even more subtle.

Sinuosity Index has a non-mathematical utility as well. Streams can be placed in categories arranged by it; for example, when the index is between 1 and 1.5 the river is sinuous, but if between 1.5 and 4, then meandering. The index is a measure also of stream velocity and sediment load, those quantities being maximized at an index of 1 (straight).

Cliffed Coast

Coastal cliffs are very steep rock faces near the sea that are greater than 5 m in height. They may ascend in steps and have ledges, crevices and overhangs. Coastal cliffs may rise directly from the sea or be separated from it by a narrow shore. Basic rocks include basalt, andesite, diorite, gabbro, and tuffaceous mudstones and sandstones. Cliffs and outcrops provide many varied habitats: from bare rock colonized only by mosses and lichens, to deeper soils supporting woody vegetation; from highly exposed situations, to heavily shaded and sheltered habitats; and from very dry to permanently wet surfaces. Coastal cliffs are particularly influenced by salt spray, with halophytes and succulents.

Cliffed Coast Morphology

Cliff Form and Occurrence

While cliffed coasts are occasionally formed in cohesionless sands where plant roots and soil moisture provide some strength, most cliffed shorelines develop in material that possesses strength due to cohesion provided by the bonding of clay minerals, cementation by chemical precipitates or the crystal bonding of igneous and metamorphic rocks. The term cliff is used here for all shorelines with a steep sub aerial slope –strictly speaking some portion of the slope should exceed 40°, the height of the cliff should exceed the maximum height of wave run-up and overtopping (though wave spray may reach the top of the cliff). If the cliff is so low that wave overtopping can occur then the shoreline feature is termed a bank. The term bluff can be used interchangeably with cliff, but here its use is restricted to describing cliffs formed in unconsolidated or not consolidated sediments, including sand, silt, clay and till. These cliffed coasts are termed cohesive coasts to distinguish them from coasts formed in much stronger bedrock where coastal processes modify the shoreline relatively slowly. While the terminology is a bit ambiguous because muddy coasts (e.g., estuarine mudflats) are also cohesive and because cohesion can equally apply to crystalline rocks, practically it is useful to make the distinction between them.

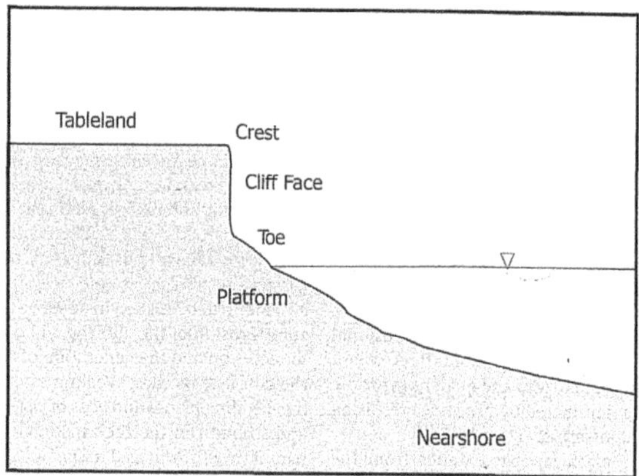

The presence of a cliffed shoreline reflects the existence of relatively high ground near the coast and this in turn may reflect tectonic forces on a continental scale, local folding and faulting, fluvial or glacial erosion, water level change, or simply recession due to coastal erosion on a gently sloping plain. Thus, the height of the coastal cliff is controlled primarily by the relief of the coastal zone and secondarily due to the operation of coastal processes.

In examining the processes and components that make up the coastal cliff recession system it is useful first of all to describe the morphological components of the system in a profile normal to the shoreline (figure) Not all features will be present on all cliffed coasts: The major components are:

a) The tableland or area inland from the cliff top.

b) The cliff top that marks the change in slope from the tableland to the cliff face and is the transition zone to the area that slopes down to the water.

c) The sub aerial difffacethat extends from the cliff top to the toe of the slope where it intersects the beach or platform. This area is dominated by erosion due to processes resulting from mass wasting as well as overland flow and gullying. In the case of plunging cliffs the face extends below the water level.

d) The cliffs toe that is the transition area between the sub aerial cliff and the beach and shore platform. The upper limit of the cliff toe is marked by height to which wave action (not including spray) can reach and the lower limit by junction with the more gently sloping shore platform. It also marks the transition from the cliff face, which is dominated by sub aerial erosional processes leading to horizontal recession, and the shore platform and nearshore profile, which are dominated by processes resulting in vertical lowering.

e) The shore platform that extends from the base of the cliff offshore to a point at, or just below spring low tide. The shore platform may be overlain 'by varying amounts of surficial sediments. The platform itself is subject to wave action as well as weathering processes during sub aerial exposure.

f) The nearshore slope that forms the subaqueous extension of the intertidal platform, and is a zone of shoaling and breald.ng waves extending offshore to the limit of wave erosion and transport of sediment.

Cliff Coast Classification

Two major types of cliffed coast can be recognised on the basis of the profile form normal to the shore and sea level (Figure below). Plunging cliffs occur where the cliff extends below the water line to some considerable depth (Figures below). Waves break directly against the cliff face, and there is no beach, ramp or platform that would lead to wave breald.ng offshore and to the accumulation of sediments. Plunging cliffs usually occur in resistant bedrock where the slope and relief are determined by tectonic events (e.g., folding or block faulting) or where erosion by glaciers or rivers, and subsequent drowning due to sea level rise, has produced a steep cliff with deep water at the cliff base. Over time, jointing and vertical weal messes may be exploited by wave action to produce chimneys and small pocket beaches, and irregularities in the cliff face may be enhanced by runoff and gully development. However, erosional processes are generally very slow on these coasts so these features form slowly. Plunging cliffs are not generally found on cohesive coasts because the cliff material is too weal< to withstand direct wave attack for very long and cliff recession soon leads to the formation of a sloping platform and beach.

Where the overall slope of the inherited coastal morphology is less steep the toe of the cliff face will be located in or above the intertidal zone. Erosion of the cliff toe will occur, leading to recession and the generation of a platform as the cliff face recedes. Weal messes such as jointing, bedding planes and beds of varying lithology and strength in the rock malting up the lower part of the cliff will lead to spatially uneven rates of erosion and the development of a variety of erosional forms such as notches, blowholes, caves, arches and stacks. The coastline tends to become highly irregular and the inner nearshore is often rocky with a variety of shallow reefs and emergent boulders (figures below). These features are absent on cohesive coasts where rapid erosion on the beach and shallow nearshore quickly removes any irregularities.

Horizontal erosion is focused at the toe of the cliff, and recession of the cliff itself will tend to produce a quasi-horizontal erosion surface or platform. However, erosional processes in the intertidal and sub-tidal zones also act on the platform leading to vertical lowering of the surface. This in turn generates a profile with an intertidal zone that slopes away from the base of the cliff and grades into the underwater nearshore profile without any abrupt transition. Type A platforms are the most common form of platform on cliffed shorelines, particularly in rocks of moderate to low strength and in areas where sand and gravel are present in the intertidal zone. However, in some areas vertical erosion of the platform in the intertidal and shallow sub-tidal zone is relatively slow compared to horizontal recession of the cliff toe. This leads to the development of a Type B shore platform that has a nearly horizontal surface away from the base of the cliff, and then terminate abruptly in a seaward drop to the nearshore (Figures below). The elevation of the platform may be close to the high-tide level, the low tide level, or somewhere in between. Type A platforms reflect conditions where vertical lowering of the platform in the intertidal area is similar to that in the inner nearshore and keeps pace with horizontal retreat of the cliff toe. The development of the nearly horizontal platform associated with type B must therefore reflect much slower vertical lowering of the intertidal platform compared to the recession of the cliff toe. The quasi-horizontal Type B platforms have generated much interest and there is considerable debate about the processes operating on them and the controls on their origin. We will examine this problem at the end of the chapter after we look at cliffed coast processes horizontal platform associated with type B must therefore reflect much slower vertical lowering of the intertidal platform compared to the recession of the cliff toe. The quasi-horizontal Type B platforms have generated much interest and there is considerable debate about the processes operating on them and the controls on their origin. We will examine this problem at the end of the chapter after we look at cliffed coast processes generally, and at processes operating on cohesive and rock coasts.

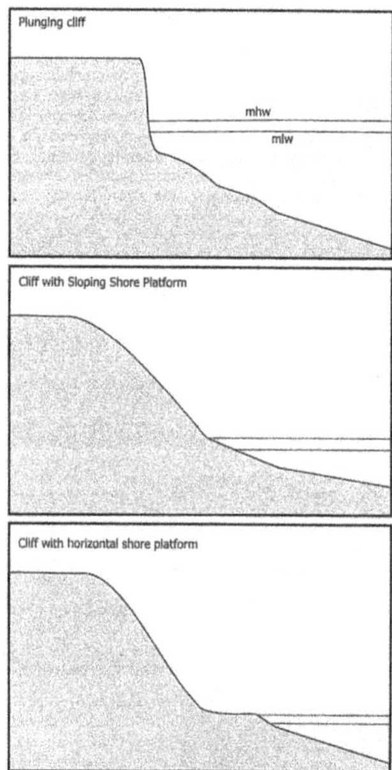

Profiles associated with the major types of cliffed coast.

Volcanic and Slope Landforms

Volcanic landforms are controlled by the geological processes that form them and act on them after they have formed. Thus, a given volcanic landform will be characteristic of the types of material it is made of, which in turn depends on the prior eruptive behavior of the volcano. Although later processes can modify the original landform, we should be able to find clues in the modified form that lead us to conclusions about the original formation process.

Shield Volcanoes

- A shield volcano is characterized by gentle upper slopes (about 50°) and somewhat steeper lower slopes (about 10°).

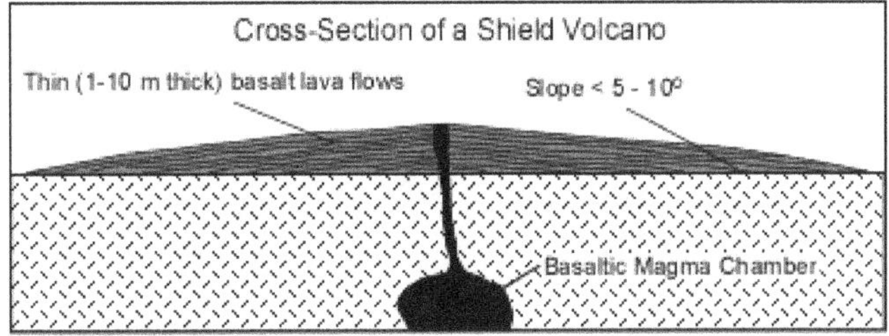

- Shield volcanoes are composed almost entirely of relatively thin lava flows built up over a central vent.
- Most shields were formed by low viscosity basaltic magma that flows easily down slope away form the summit vent.
- The low viscosity of the magma allows the lava to travel down slope on a gentle slope, but as it cools and its viscosity increases, its thickness builds up on the lower slopes giving a somewhat steeper lower slope.
- Most shield volcanoes have a roughly circular or oval shape in map view.
- Very little pyroclastic material is found within a shield volcano, except near the eruptive vents, where small amounts of pyroclastic material accumulate as a result of fire fountaining events.
- Shield volcanoes thus form by relatively non-explosive eruptions of low viscosity basaltic magma.

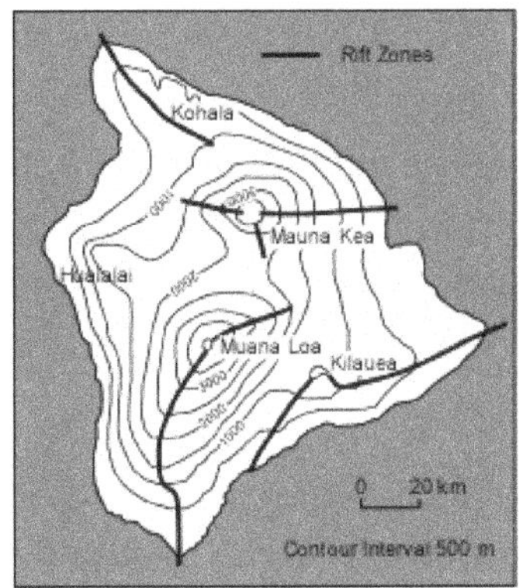

- Vents for most shield volcanoes are central vents, which are circular vents near the summit. Hawaiian shield volcanoes also have flank vents, which radiate from the summit and take the form of en-echelon fractures or fissures, called rift zones, from which lava flows are emitted. This gives Hawaiian shield volcanoes like Kilauea and Mauna Loa their characteristic oval shape in map view.

Stratovolcanoes

- Have steeper slopes than shield volcanoes, with slopes of 6 to 100 low on the flanks to 30° near the top.
- The steep slope near the summit is due partly to thick, short viscous lava flows that do not travel far down slope from the vent.

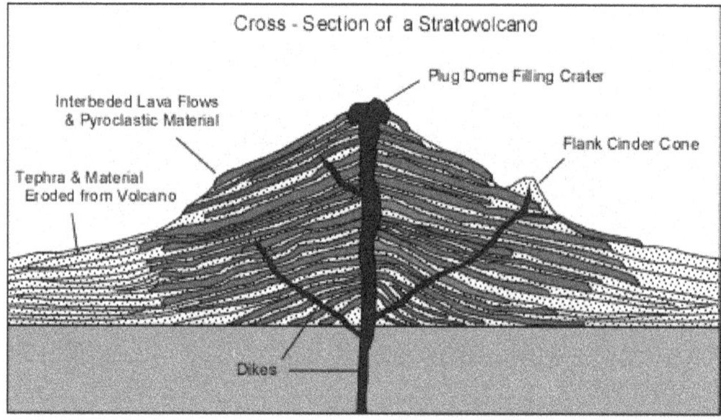

- The gentler slopes near the base are due to accumulations of material eroded from the volcano and to the accumulation of pyroclastic material.
- Stratovolcanoes show inter-layering of lava flows and pyroclastic material, which is why they are sometimes called composite volcanoes. Pyroclastic material can make up over 50% of the volume of a stratovolcano.
- Lavas and pyroclastics are usually andesitic to rhyolitic in composition.
- Due to the higher viscosity of magmas erupted from these volcanoes; they are usually more explosive than shield volcanoes.
- Stratovolcanoes sometimes have a crater at the summit that is formed by explosive ejection of material from a central vent. Sometimes the craters have been filled in by lava flows or lava domes, sometimes they are filled with glacial ice, and less commonly they are filled with water.
- Long periods of repose (times of inactivity) lasting for hundreds to thousands of years, make this type of volcano particularly dangerous, since many times they have shown no historic activity, and people are reluctant to heed warnings about possible eruptions.

Maars

- Maars result from phreatic or phreatomagmatic activity, wherein magma heats up groundwater, pressure builds as the water to turns to steam, and then the water and preexisting rock (and some new magma if the eruption is phreatomagmatic) are blasted out of the ground to form a tephra cone with gentle slopes.

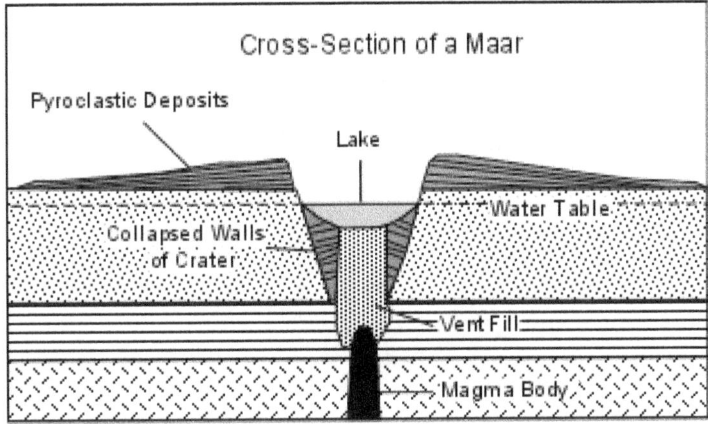

Parts of the crater walls eventually collapse back into the crater, the vent is filled with loose material, and, if the crater still is deeper than the water table, the crater fills with water to form a lake, the lake level coinciding with the water table.

Volcanic Domes

- Volcanic Domes result from the extrusion of highly viscous, gas poor andesitic and rhyolitic lava. Since the viscosity is so high, the lava does not flow away from the vent, but instead piles up over the vent.
- Blocks of nearly solid lava break off the outer surface of the dome and roll down its flanks to form a breccia around the margins of domes.
- The surface of volcanic domes are generally very rough, with numerous spines that have been pushed up by the magma from below.

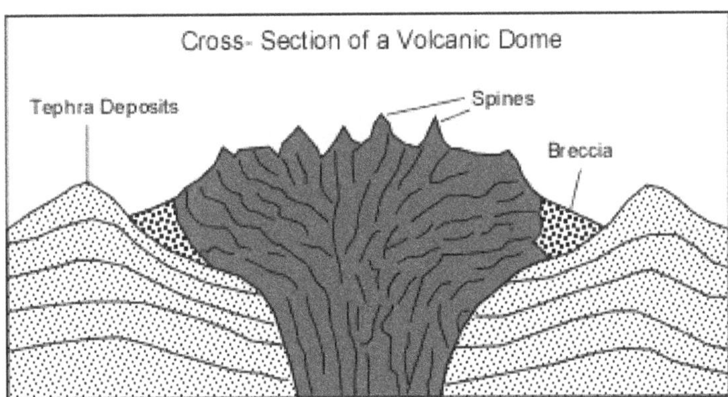

- Most dome eruptions are preceded by explosive eruptions of more gas rich magma, producing a tephra cone into which the dome is extruded.
- Volcanic domes can be extremely dangerous. Because they form unstable slopes that may collapse to expose gas-rich viscous magma to atmospheric pressure. This can result in lateral blasts or Pelean type pyroclastic flow eruptions.

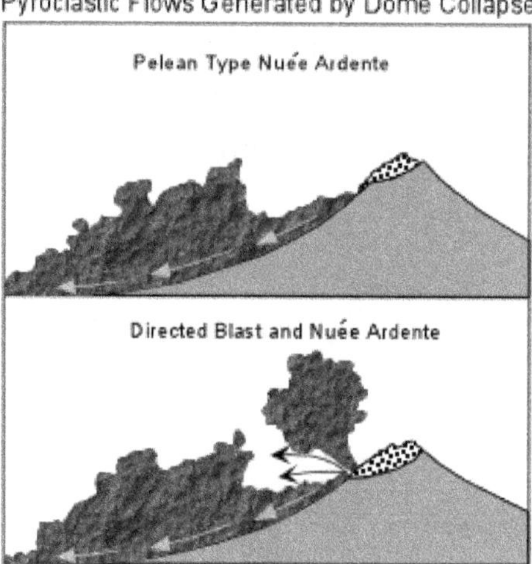

Resurgent Domes

- After the formation of a caldera by collapse, magma is sometimes re-injected into the area below the caldera. This can result in uplift of one or more areas within the caldera to form a resurgent dome. Two such resurgent domes formed in the Yellowstone caldera.
- If magma leaks back to the surface during this resurgent doming, then eruptions of small volcanic domes can occur in the area of the resurgent domes.

Geysers, Fumaroles and Hot Springs

- A fumarole is vent where gases, either from a magma body at depth, or steam from heated groundwater, emerges at the surface of the Earth. Since most magmatic gas is H_2O vapor, and since heated groundwater will produce H_2O vapor, fumaroles will only be visible if the water condenses. (H_2O vapor is invisible, unless droplets of liquid water have condensed).

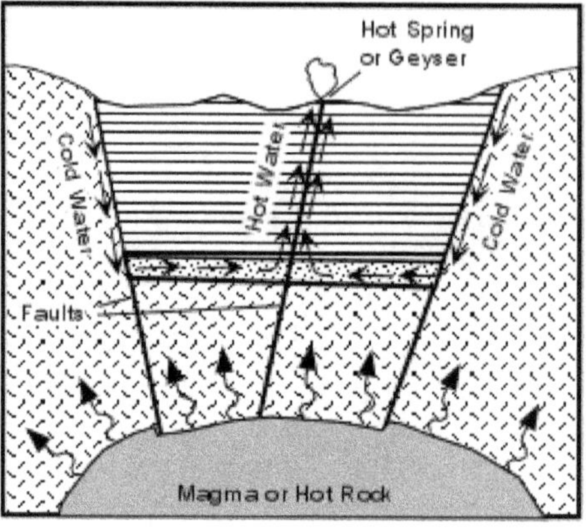

After Williams and McBirney, 1985

- Hot springs or thermal springs are areas where hot water comes to the surface of the Earth. Cool groundwater moves downward and is heated by a body of magma or hot rock. A hot spring results if this hot water can find its way back to the surface, usually along fault zones.

Minerals dissolved in the high temperature water are often precipitated when the water cools at the surface. This produces spectacular deposits of travertine (chemically precipitated calcite, or siliceous sinter.

Bacteria forming microbial mats under the water are responsible for the coloration often seen in hot springs. Different species, with different colors thrive at different temperatures.

- A geyser results if the hot spring has a plumbing system that allows for the accumulation of steam from the boiling water. When the steam pressure builds so that it is higher than the pressure of the overlying water in the system, the steam will move rapidly toward the surface, causing the eruption of the overlying water. Some geysers, like Old Faithful in Yellowstone Park, erupt at regular intervals. The time between eruptions is controlled by the time it takes for the steam pressure to build in the underlying plumbing system.

Volcanic Cone

A volcanic cone is a triangle-shaped hill formed as material from volcanic eruptions piles up around the volcanic vent, or opening in Earth's crust.

Most volcanic cones have one volcanic crater, or central depression, at the top. They are probably the most familiar type of volcanic mountain.

Major Types of Volcanic Cones

Composite Cones

Composite cones are some of the most easily recognizable and imposing volcanic mountains, with sloping peaks rising several thousand meters above the landscape.

Also known as stratocones, composite cones are made up of layers of lava, volcanic ash, and fragmented rocks. These layers are built up over time as the volcano erupts through a vent or group of vents at the summit's crater. The eruptions that form these cones, called Plinian eruptions, are violently explosive and often dangerous.

One of the most famous stratocones in the world is Mount Fuji, Japan. The tallest mountain in Japan, Mount Fuji towers 3,776 meters (12,380 feet) above the surrounding landscape. Mount Fuji last erupted in 1707, but is still considered an active volcano.

Mount Rainier, Washington, is another stratocone. Mount Rainier rises 4,392 meters (14,410 feet) above sea level. Over the past half million years, Mt. Rainier has produced a series of alternating lava eruptions and debris eruptions. These eruptions have given Mt. Rainier the classic layered structure and sloping shape of a composite cone. Unlike Mount Fuji, Mount Rainier's composite cone has been carved down by a series of glaciers, giving it a craggy and rugged shape.

Cinder Cones

Cinder cones, sometimes called scoria cones or pyroclastic cones, are the most common types of volcanic cones. They form after violent eruptions blow lava fragments into the air, which then solidify and fall as cinders around the volcanic vent. Usually the size of gravel, these cinders are filled with many tiny bubbles trapped in the lava as it solidifies. Cinder cones stand at heights of tens of meters to hundreds of meters.

Cinder cones may form by themselves or when new vents open on larger, existing volcanoes. Mauna Kea, a volcano on the American island of Hawaii, and Mount Etna, a volcano on the Italian island of Sicily, are both covered with hundreds of cinder cones.

Other Types of Volcanic Cones

Spatter Cones

Volcanoes often eject small amounts of gaseous lava blobs into the air. These lava blobs, called spatter, are heavy and viscous. Viscosity refers to a substance's resistance to flow. In this case, it refers to the spatter's thickness. The viscosity of spatter means it often does not have time to cool before hitting the ground.

The lava blobs in spatter stick together as they land, piling up to form steep-sided spatter cones. Most spatter cones are very small, ranging between 1 and 5 meters (3 to 16 feet) in height, because they result from minor volcanic activity. They often form in linear groups along an eruptive fissure, or long crack, on the flank of an active volcano. A small spatter cone is called a hornito.

Spatter cones can be found in and around the Pu'u 'Ō'ō region of Mount Kilauea in Hawaii. Continuously erupting since 1983, Kilauea's volcanic activity is characterized by the fountaining of hot lava, making it the perfect incubator for spatter cones.

Tuff Cones

Unlike spatter cones that form from lava fountaining, tuff cones form from the interaction between rising magma and bodies of water. Tuff cones are sometimes called ash cones.

When heated rapidly by lava, water flashes to steam and expands violently, fragmenting huge amounts of lava into plumes of very fine grains of ash. This ash falls around the volcanic vent, creating an ash cone. Over time, the ash weathers into a rock known as tuff.

Tuff cones have steep sides and often stand between 100 and 300 meters (328 to 984 feet) high. They are much wider and have broader craters than spatter cones because they result from shallow explosions that eject materials sideways rather than upwards.

Diamond Head, the famous volcano near Honolulu, Hawaii, is an enormous tuff cone. The mountain is the result of a brief volcanic eruption about 200,000 years ago. During Diamond Head's eruption period, the mountain rose from the ocean, and lava interacted with water and even a nearby coral reef. Today, Diamond Head's rim is about a kilometer (.62 mile) from the coast, and rises about 232 meters (760 feet) above sea level.

Wizard Island is a small cinder cone located inside the remains of a much larger volcano, Mount Mazama. Mount Mazama was a composite cone volcano that collapsed about 7,700 years ago and formed Crater Lake, Oregon.

Volcanic Crater

A volcanic crater is a roughly circular depression in the ground caused by volcanic activity. It is typically a bowl-shaped feature within which occurs a vent or vents. During volcanic eruptions, molten magma and volcanic gases rise from an underground magma chamber, through a tube-shaped conduit, until they reach the crater's vent, from where the gases escape into the atmosphere and the magma is erupted as lava. A volcanic crater can be of large dimensions, and sometimes of great depth. During certain types of explosive eruptions, a volcano's magma chamber may empty enough for an area above it to subside, forming a type of larger depression known as a caldera.

Craters on Mount Cameroon

Geomorphology

In most volcanoes, the crater is situated at the top of a mountain formed from the erupted volcanic deposits such as lava flows and tephra. Volcanoes that terminate in such a summit crater are usually of a conical form. Other volcanic craters may be found on the flanks of volcanoes, and these are commonly referred to as flank craters. Some volcanic craters may fill either fully or partially with rain and/or melted snow, forming a crater lake.

A crater may be breached during an eruption, either by explosions or by lava, or through later erosion. Breached craters have a much lower rim on one side.

Some volcanoes, such as maars, consist of a crater alone, with scarcely any mountain at all. These volcanic explosion craters are formed when magma rises through water-saturated rocks, which causes a phreatic eruption. Volcanic craters from phreatic eruptions often occur on plains away from other obvious volcanoes. Not all volcanoes form craters.

The volcanic crater of a Tangkuban Parahu mount, Bandung, Indonesia

Plateau

Plateaus can be formed by the erosion of the surrounding areas by rivers, flooding, and glacier. Repeated lava flows can also form a plateau over a period of time. Plateau is formed when the magma pushes up towards the surface of the earth's crust. There are many ways in which plateaus can develop. The magma does not break through but instead raises a section of the crust up as it raises and creates a plateau. It is also formed by the effects of that millions of years of wind and water erosion can have on land areas. Water that flows in specific pattern layers the rock at time and sculptures the channels that will continuously dig deeper. As the water digs its way down, steep slopes are created and that forms the edges of the plateau. Plateaus are very useful because they are rich in mineral deposits. As a result, many of the mining areas in the world are located in the plateau areas.

Types of the Plateau

- Plateaus are divided into three types
- It is based on location, where they are situated.
    1. Intermontane plateaus
    2. Piedmont plateaus
    3. Continental plateaus

Volcanic Plateau

- It is formed by many small volcanic eruptions which slowly build up by time which forms plateau resulting in lava flows. Example- Deccan traps.

Tibetan Plateau

- It is called the roof of the world because it is the highest and largest plateau in the world.
- It is formed due to the collision of the Indo-Australian and Eurasian tectonic plates.
- The Himalayan range of the south mountain is surrounded by it.

Columbia –Snake Plateau

- It is formed by the volcanic eruptions with a related coating of basalt lava.
- It meets the river Columbia and its tributary snake meets in this plateau.

Deccan Plateau

- It is the plateau which forms the southern part of India and is the largest plateau.

- The Western Ghats and the Eastern Ghats are bordered by Deccan plateau.
- This is also known for containing some unique fossils.

Katanga Plateau

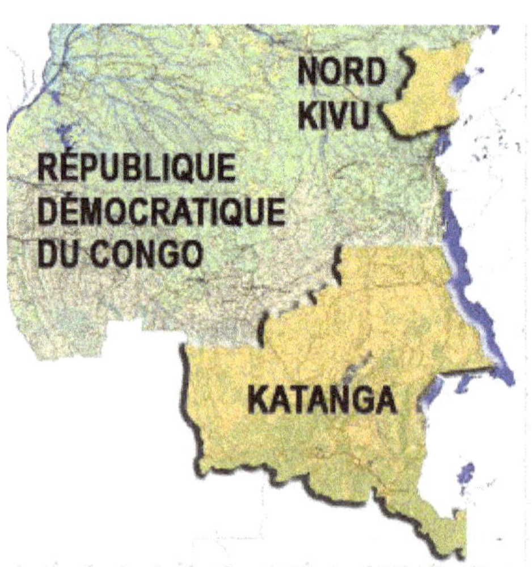

- Katanga plateau is famous for copper production.
- This lies in Congo.
- Minerals found here is zinc, silver, gold, and cobalt are mined here.

Facts about Plateau

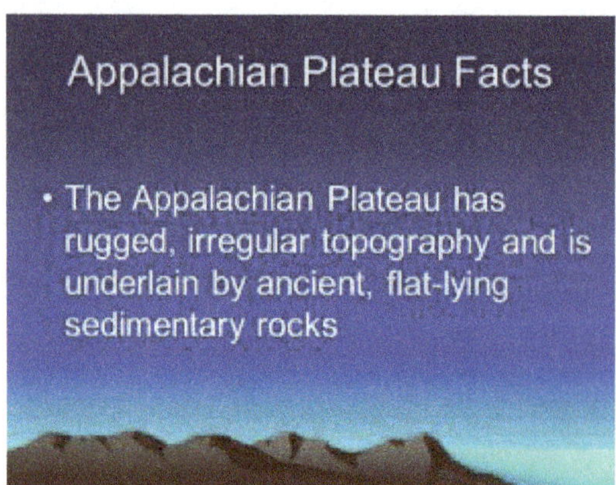

- Plateaus can be formed by a number of processes like upwelling of volcanic magma, extrusion of lava, and erosion by water and glaciers.
- Plateaus can also be built up lava spreading outward from cracks and weak areas in the crust.
- Plateaus can also be formed by the erosional processes of glaciers on mountain ranges, leaving them sitting between the mountain ranges.
- Piedmont plateaus are bordered on one side by mountains and on the other by plain or sea

Plains

A plain is a type of landform made up of a flat area that can exist in valleys, lowlands, on plateaus, or uplands. They are formed by a variety of weather and geological phenomena including water deposits, ice, wind, erosion, and even lava. Plains make excellent agricultural ground in many places because of their rich soil and their relatively flat landscape. There are different types of plains including structural plains, erosional plains, and depositional plains, as well as coastal plains and flood plains. Natural plant life on plains is dependent on the climate and can range from thick forests, to grasslands, and everything in between.

Interesting Plains Facts

- Structural plains tend to be large flat surfaces that make up extensive lowlands.
- Erosional plains are those that have been created by erosion die to glaciers, wind, running water and rivers.
- Depositional plains are created when material is deposited from rivers, glaciers, waves and wind. Sometimes these plains are very fertile because of the type of material that has been deposited there.
- Depositional plains are classified as alluvial plains, or glacial plains.
- Alluvial plains are created by a river that deposits material that becomes the soil.
- Flood plains are plains that experience periodic flooding or just occasional flooding. One of the most famous flood plains is the one surrounding the Nile River in Africa.
- A lacustrine plain is a plain that was originally the bottom of a lake.
- A lava plain is formed when lava creates sheets over time, eventually becoming rich soil.
- Glacial plains are formed when a glacier moves across land and the force of gravity creates the large flat surface.
- An abyssal plain is an area of the ocean basin that is either flat or sloping very gently.
- Plains that exist on other planets are referred to as planitia (Latin for the word plain). Mars has Hellas Planitia and Venus has Sedna Planitia.
- The Great Plains in North America cover areas of ten states including New Mexico, Texas, Oklahoma, Colorado, Kansas, Wyoming, Nebraska, South Dakota, North Dakota, and Montana. Weather on the Great Plains can be extreme at times, which has led to a lot of irrigation to help compensate for drought conditions that can occur.
- Coastal plains are lowlands that stretch along the shore and slope towards it. An example of a coastal plain is the Atlantic Coastal Plain. This plain stretches from Florida to Nova Scotia.
- Many of the world's rivers are surrounded by plains.
- Plains cover approximately one-third of the land on earth.

- Every continent on earth has plains in one form or another.
- In North America, where some plains are grasslands, these are referred to as prairies. These types of plains have warm summers and cold winters.
- Mexico has a forested plain called Tabasco Plain that is home to all sorts of vegetation including trees and shrubs.
- A tropical grassland plain is called a savannah. Savannahs are warm all year and tend to have scattered trees. Africa's main savannah is the Serengeti.
- In very cold climates like the Arctic, plains are often frozen. They are referred to as tundra.

References

- Neuendorf, K.E.K., J.P. Mehl Jr., and J.A. Jackson, 2005, Glossary of Geology. Springer-Verlag, New York City, 779 pp., ISBN 3-540-27951-2
- What-are-landforms-and-major-types-of-landforms-on-earth: eartheclipse.com, Retrieved 29 July 2018
- Kukla, G. J. (1977). "Pleistocene Land-Sea Correlations I. Europe". Earth-Science Reviews. 13: 307–374. Bibcode:1977ESRv...13..307K. doi:10.1016/0012-8252(77)90125-8
- Aeolian-landforms-what-is-a-yardang: worldatlas.com, Retrieved 09 May 2018
- Bowker, Kent A. (1988). "Albert Einstein and Meandering Rivers". Earth Science History. 1 (1). Retrieved 2016-07-01
- Mountain-and-glacial-landforms-what-is-a-cirque: worldatlas.com, Retrieved 15 March 2018
- Thiel, C.; Buylaert, J. P.; Murray, A. S.; Terhorst, B.; Tsukamoto, S.; Frechen, M.; Sprafke, T. (2011). "Investigating the chronostratigraphy of prominent palaeosols in Lower Austria using post-IR IRSL dating". Quaternary Science Journal. 60 (1): 137–152. doi:10.3285/eg.60.1.10
- Valley, science: britannica.com, Retrieved 19 June 2018
- Getis, Arthur; Judith Getis and Jerome D. Fellmann (2000). Introduction to Geography, Seventh Edition. McGraw Hill. p. 99. ISBN 0-697-38506-X
- The-glacial-landforms-and-cycle-of-erosion-744: geographynotes.com, Retrieved 11 April 2018
- Murray, A. S.; Wintle, A. G. (2000). "Luminescence dating of quartz using an improved single aliquot regenerative-dose protocol". Radiation Measurements. 32: 57–73. Bibcode:2000RadM...32...57M. doi:10.1016/S1350-4487(99)00253-X
- Coastal-landform, science: britannica.com, Retrieved 27 May 2018
- Schmidt, E. D.; Semmel, A.; Frechen, M. (2011). "Luminescence dating of the loess/palaeosol sequence at the gravel quarry Gaul/Weilbach, Southern Hesse (Germany)". Quaternary Science Journal. 60 (1): 116–125. doi:10.3285/eg.60.1.08
- Geography-of-river-deltas-1435824: thoughtco.com, Retrieved 30 June 2018

5

Erosion and Deposition

The action of surface processes that transports dissolved material, soil or rock from one location to another is known as erosion. Deposition is another geological process that adds sediments, rocks and soil to a land mass. It is caused due to wind, water, ice or gravity transport. This chapter explores the different types of erosion and deposition processes such as coastal, internal, soil, sheet, glacial erosion, etc.

Erosion

Erosion is the geological process in which earthen materials are worn away and transported by natural forces such as wind or water. A similar process, weathering, breaks down or dissolves rock, but does not involve movement.

Erosion is the opposite of deposition, the geological process in which earthen materials are deposited, or built up, on a landform.

Most erosion is performed by liquid water, wind, or ice (usually in the form of a glacier). If wind is dusty, or water or glacial ice is muddy, erosion is taking place. The brown color indicates that bits of rock and soil are suspended in the fluid (air or water) and being transported from one place to another. This transported material is called sediment.

Physical and Chemical Erosion

The process of erosion is often broken down into two forms: physical erosion and chemical erosion. They often work together, as well as with other geological processes such as weathering and sedimentation.

Physical Erosion

Physical erosion describes the process of rocks changing their physical properties without changing their basic chemical composition. Physical erosion often causes rocks to get smaller or smoother. Rocks eroded through physical erosion often form clastic sediments. Clastic sediments are composed of fragments of older rocks that have been transported from their place of origin.

Landslides and other forms of mass wasting are associated with physical weathering. These processes cause rocks to dislodge from hillsides and crumble as they tumble down a slope.

Plant growth can also contribute to physical erosion in a process called bio erosion. Plants break up earthen materials as they take root, and can create cracks and crevices in rocks they encounter.

Ice and liquid water can also contribute to physical erosion as their movement forces rocks to crash together or crack apart. Some rocks shatter and crumble, while others are worn away. River rocks are often much smoother than rocks found elsewhere, for instance, because they have been eroded by constant contact with other river rocks.

Chemical Erosion

Chemical erosion describes the process of rocks changing their chemical composition as they erode. Chemical erosion almost always refers to rocks interacting and undergoing a chemical reaction with water.

The most familiar form of chemical erosion is probably rust, the product of a process called oxidation. During oxidation, rocks interact with oxygen in the presence of water. The amount of water required for oxidation is minimal, often the amount of water present in the atmosphere. Iron is the most familiar mineral to undergo oxidation and rust.

Carbonation is another form of chemical erosion. During carbonation, rocks interact with carbon dioxide in the presence of water. In rocks such as chalk, carbonation can create a weak acid (carbonic acid) that erodes the surface of the rock.

Hydration is a form of chemical erosion in which the chemical bonds of the mineral are changed as it interacts with water. One instance of hydration occurs as the mineral anhydrite reacts with groundwater. The water transforms anhydrite into gypsum, one of the most common minerals on Earth.

Another familiar form of chemical erosion is hydrolysis. In the process of hydrolysis, a new solution (a mixture of two or more substances) is formed as chemicals in rock interact with water. In many rocks, for example, sodium minerals interact with water to form a saltwater solution.

Various Causes of Erosion

Water

Water is a liquid drank by humans, animals, and plants. It's also produced during photosynthesis. Rain and subsequent flowing water carry along weathered rocks and other particles to be deposited at lower elevations.

Wind

Wind is the movement of air, in most cases with significant force. The movement of air is usually from an area of high pressure to an area of low pressure. The force of wind is capable of carrying eroded materials to different locations.

Ice

Ice forms during periods when temperatures considerably drop, resulting in the pile-up of snow and ice. The result is continental ice sheets. Ice sheets erode mountain surfaces over time to form beautiful landforms. The movement of ice downhill causes erosion of underlying rocks, leading to nicely carved up landscapes.

Gravity

This is the force of attraction between two objects. The force of attraction relies on the masses of the two objects and the distance between them. Gravitational force is responsible for the downward movement of water and particles.

Waves

Waves are moving swells or ridges in a water body. Kinetic energy is transmitted in the direction of the wave movement. The energy is capable of carrying away sand particles to different locations.

Effects of Erosion

Reduction of Soil Fertility

Repeated erosion washes away the topsoil. The top soil is loaded with nutrients and organic matter critical for crop growth. Extensive erosion also minimizes the depth of soil available for water storage and rooting. Repeated erosion reduces water infiltration into the soil, which may result to withering of crops. Erosion also enhances run off, which create unsightly gullies.

Damage and Increased Costs

Erosion leads to massive deposition of sediments on roads, and railways. This may cut off transportation lines. Costs will be incurred in regards to clearing away the deposition on the transport lines to allow transportation to resume.

Erosions may also culminate to landslides, which can damage buildings and cause deaths to people living at the foot of the hills or mountains. The cost incurred in the rehabilitation of an area after a landslide is substantial.

Erosion can lead to loss of pesticides, fertilizers, seeds, and seedlings. It also necessitates repeat of field operations. Repeating field operations mean more expenses are added to previous failed operations.

Young plants may fail to make it to maturity due to being blasted by particles flown around by wind. An extra cost of cultivation will be incurred due to the need to level up eroded surfaces.

Environmental Impacts

Erosion leads to huge deposition of sediments into drains. This may cause drainage problems. Water sources such as rivers, streams, and lakes can be polluted through extensive inputs of pesticides, nitrogen and phosphorous.

Deposition of sediments in rivers can damage river ecosystems. Deposited sediment pollutes rivers and cuts off oxygen supplies leading to fish deaths. Deposited sediment in rivers can also cut off supply of fresh water to people living downstream. These communities would be forced to find alternative sources of water or walk distances in search of this precious commodity. Increased deposition of sediment and runoff can lead to massive flooding downstream. Flooding can destroy property and cause deaths. Floods are fertile breeding grounds for mosquitoes. This might result in the upsurge of malaria in the area.

Water Erosion

Water plays a significant role in rock erosion since it's able to move these weathered materials from one point to another. Moving water such as currents in oceans or rivers plays a significant role in erosion because they move materials from their primary source to a separate location. Erosion may discolor rivers as they snake through the valleys to oceans or seas. This is due to the huge amount of sediment deposited by the process of erosion. Once these eroded materials are settled and piled up in a new location, it is referred to as deposition. Water is also able to erode land by the effects of currents and ocean waves. Once the eroded particles as a result of ocean currents and waves are settled and deposited, they enormously change the coastline of the area.

Wind Erosion

Wind blows away weathered particles from the source to other locations. Wind can also speed up the erosional capability of water. For instance, when a raindrop is released from the sky, it's relatively weak. The force of the wind gives it more momentum such that when it hits the surface of the earth, it able to carry away a significant amount of particles.

Due to the effect of wind, the raindrop can travel at a speed of 32km/hr. At this breakneck speed, it's able to steadily break down rock material and soil, and make erosion and transportation a lot easier. The effect of wind is usually manifested in areas that experience less or no rain or dry and barren land that is not capable of supporting vegetation.

A typical example of this phenomenon is the Middle East dust bowls that took place in the course of the great depression. Wind causes erosion of rock particles driven by soil and sand particles that are not tightly glued together and not insulated by vegetation. The carrying away of dry soil and loose sand particles is known as deflation. The action of wind continues until that time when the power and momentum of wind cannot move the loose particles.

Glacier Erosion

Glacier is an enormous sheet of snow-covered ice that slowly accumulates on a mountain. When the ice below it starts to melt, the glacier may start to move, consequently, eroding the mountain. Glaciers form in areas that are frequently covered by snow. The amount of snow falling each year usually outweighs the amount that melts, resulting in massive accumulation of snow.

When snow accumulates, the snow above exerts a lot of pressure on the snow below, which triggers it to recrystallize and transform into solid ice. As the glacier moves across the landscape, it picks up almost everything in their path including sand grains and giant boulders. As these sand grains and giant boulders get hauled over across the bedrock, they act as cutting tools carving out the bedrock as the glacier moves. This pretty much explains how glacier causes erosion.

Coastal Erosion

Coastal erosion is the wearing away of the land by the sea. This often involves destructive waves wearing away the coast (though constructive waves also contribute to coastal erosion).

Causes of Coastal Erosion

Coastal erosion is typically driven by the action of waves and currents, but also by mass wasting processes on slopes, and subsidence (particularly on muddy coasts). Significant episodes of coastal erosion are often associated with extreme weather events (coastal storms, surge and flooding) but also from tsunami, both because the waves and currents tend to have greater intensity and because the associated storm surge or tsunami inundation can allow waves and currents to attack landforms which are normally out of their reach. On coastal headlands, such processes can lead to undercutting of cliffs and steep slopes and contribute to mass wasting. In addition, heavy rainfall can enhance the saturation of soils, with high saturation leading to a reduction in the soil's shear strength, and a corresponding increase in the chance of slope failure (landslides).

Coastal erosion is a natural process which occurs whenever the transport of material away from the shoreline is not balanced by new material being deposited onto the shoreline. Many coastal landforms naturally undergo quasi-periodic cycles of erosion and accretion on time-scales of days to years. This is especially evident on sandy landforms such as beaches, dunes, and intermittently closed and open lagoon entrances. However, human activities can also strongly influence the propensity of landforms to erode. For example, the construction of coastal structures (such as breakwaters, groynes and seawalls) can lead to changes in coastal sediment transport pathways, resulting in erosion in some areas and accretion in others. The removal of sediments from the coastal system (e.g., by dredging or sand mining), or a reduction in the supply of sediments (e.g., by the regulation of rivers) can also be associated with unintended erosion. At larger scales, natural and human-induced climate change can modulate the likelihood and rate of coastal erosion.

Coastal erosion becomes a hazard when society does not adapt to its effects on people, the built environment and infrastructure.

Small-scale erosion destroys abandoned railroad tracks

A place where erosion of a cliffed coast has occurred is at Wamberal in the Central Coast region of New South Wales where houses built on top of the cliffs began to collapse into the sea. This is due to waves causing erosion of the primarily sedimentary material on which the buildings foundations sit.

Dunwich, the capital of the English medieval wool trade, disappeared over the period of a few centuries due to redistribution of sediment by waves. Human interference can also increase coastal erosion: Hallsands in Devon, England, was a coastal village that washed away over the course of a year, an event directly caused by dredging of shingle in the bay in front of it.

The California coast, which has soft cliffs of sedimentary rock and is heavily populated, regularly has incidents of housing damage as cliffs erode. Devil's Slide, Santa Barbara, the coast just north of Ensenada, and Malibu are regularly affected.

The Holderness coastline on the east coast of England, just north of the Humber Estuary, is one of the fastest eroding coastlines in Europe due to its soft clay cliffs and powerful waves. Groynes and other artificial measures to keep it under control has only accelerated the process further down the coast, because long shore drift starves the beaches of sand, leaving them more exposed. The White Cliffs of Dover have also been affected.

Fort Ricasoli in Kalkara, Malta already showing signs of damage where the land is being eroded.

Fort Ricasoli, a historic 17th century fortress in Malta is being threatened by coastal erosion, as it was built on a fault in the headland, which is prone to erosion. A small part of one of the bastion walls has already collapsed since the land under it has eroded, and there are cracks in other walls as well.

Control Methods

There are three common forms of coastal erosion control methods. These three include: soft-erosion controls, hard-erosion controls, and relocation.

Hard-erosion Controls

Hard-erosion control methods provide a more permanent solution than soft-erosion control methods. Seawalls and groynes serve as semi-permanent infrastructure. These structures are not immune from normal wear-and-tear and will have to be refurbished or rebuilt. It is estimated the average life span of a seawall is 50–100 years and the average for a groyne is 30–40 years. Because of their relative permanence, it is assumed that these structures can be a final solution to erosion. Seawalls can also deprive public access to the beach and drastically alter the natural state of the beach. Groynes also drastically alter the natural state of the beach. Some claim that groynes could reduce the interval between beach nourishment projects though they are not seen as a solution to beach nourishment. Other criticisms of seawalls are that they can be expensive, difficult to maintain, and can sometimes cause further damage to the beach if built improperly.

This image represents a typical seawall that is used for preventing and controlling coastal erosion.

Natural forms of hard-erosion control include planting or maintaining native vegetation, such as mangrove forests and coral reefs.

Soft-erosion Controls

Sand bagged beach at the site of Hurricane Sandy.

Soft erosion strategies refer to temporary options of slowing the effects of erosion. These options, including Sandbag and beach nourishment, are not intended to be long term solutions or permanent solutions. Another method, beach scraping or beach bulldozing allows for the creation of an artificial dune in front of a building or as means of preserving a building foundation. However, there is a U.S. federal moratorium on beach bulldozing during turtle nesting season, 1 May – 15 November. One of the most common methods of soft erosion control is beach nourishment projects. These projects involve dredging sand and moving it to the beaches as a means of reestablishing the sand lost due to erosion. In some situations, beach nourishment is not a suitable measure to take for erosion control, such as in areas with sand sinks or frequent and large storms.

Relocation

Under this response, humans move from the coast and surrender the coast to the natural processes of both absolute and relative sea level rise and erosion. This solution is eco-centric meaning that the focus is on forcing humans to adapt to the natural processes rather than the opposite. By removing structures along the oceanfront, the beach is surrendered to the natural forces of the ocean. In this case, property owners and coastal communities are essentially "retreating" from the sea. Depending on factors such as the severity of the erosion, as well as the natural landscape of the property, relocation could simply mean moving inland by a number of yards, or in more severe cases, relocation can be to completely desert an area. Typically, there has been low public support for "retreating." However, this would be most effective in reducing the impacts of erosion on human society.

Coastal Processes

There are five main processes of coastal erosion.

Corrasion

Corrasion is a from of erosion in a riverbed or seabed where small particles are removed by motion in the water. This process occurs when high energy waves have the energy to be able to carry pebbles with force. As the wave breaks at the foot of the cliff, material is thrown at the cliff face and wears it away by chipping fragments off.

Corrosion

Corrosion mainly takes place on limestone cliffs, such as those around Dover or Beachy Head, when the acidic sea water dissolves the chalk. It can also occur by the evaporation of salt, forming crystals that expand and cause the rock to disintegrate.

Corrosion is a dangerous and extremely costly problem. Because of it, buildings and bridges can collapse, oil pipelines break, chemical plants leak, and bathrooms flood. Corroded electrical contacts can cause fires and other problems, corroded medical implants may lead to blood poisoning, and air pollution has caused corrosion damage to works of art around the world. Corrosion threatens the safe disposal of radioactive waste that must be stored in containers for tens of thousands of years.

Corrosion is one of the most damaging and costly naturally occurring events.

The most common kinds of corrosion result from electrochemical reactions. General corrosion occurs when most or all of the atoms on the same metal surface are oxidized, damaging the entire surface. Most metals are easily oxidized: they tend to lose electrons to oxygen (and other substances) in the air or in water. As oxygen is reduced (gains electrons), it forms an oxide with the metal.

When reduction and oxidation take place on different kinds of metal in contact with one another, the process is called galvanic corrosion. In electrolytic corrosion, which occurs most commonly in electronic equipment, water or other moisture becomes trapped between two electrical contacts that have an electrical voltage applied between them. The result is an unintended electrolytic cell.

Take a metal structure such as the Statue of Liberty. It looks strong and permanent. Like nearly all metal objects, however, it can become unstable as it reacts with substances in its environment and deteriorates. Sometimes this corrosion is harmless or even beneficial: the greenish patina that covers the statue's copper skin protected the metal beneath from weather damage. Inside the statue, however, corrosion caused serious harm over the years. Its iron frame and copper skin acted like the electrodes of a huge galvanic cell, so that nearly half of the frame had rusted away by 1986, the statue's one hundredth anniversary.

Natural Protection

Some metals acquire a natural passivity, or resistance to corrosion. This occurs when the metal reacts with, or corrodes in, the oxygen in air. The result is a thin oxide film that blocks the metal's tendency to undergo further reaction. The patina that forms on copper and the weathering of certain sculpture materials are examples of this. The protection fails if the thin film is damaged or destroyed by structural stress — on a bridge, for example — or by scraping or scratching. In such cases the material may repassivate, but if that is not possible, only parts of the object corrode. Then the damage is often worse because it is concentrated at these sites.

Harmful corrosion can be prevented in numerous ways. Electrical currents can produce passive films on metals that do not normally have them. Some metals are more stable in particular environments than others, and scientists have invented alloys such as stainless steel to improve performance under particular conditions. Some metals can be treated with lasers to give them a non-crystalline structure, which resists corrosion. In galvanization, iron or steel is coated with the more active zinc; this forms a galvanic cell where the zinc corrodes rather than the iron. Other metals are protected by electroplating with an inert or passivating metal. Non-metallic coatings — plastics, paints, and oils — can also prevent corrosion.

Hydraulic Action

Hydraulic action is the erosion that occurs when the motion of water against a rock surface produces mechanical weathering. Most generally, it is the ability of moving water (flowing or waves) to dislodge and transport rock particles. Within this rubric are a number of specific erosional processes, including abrasion, attrition, corrasion, saltation, and scouring (downcutting). Hydraulic action is distinguished from other types of water facilitated erosion, such as *static erosion* where water leaches salts and floats off organic material from unconsolidated sediments, and from *chemical erosion* more often called chemical weathering. It is a mechanical process, in which the moving water current flows against the banks and bed of a river, thereby removing rock particles.

A primary example of hydraulic action is a wave striking a cliff face which compresses the air in cracks of the rocks. This exerts pressure on the surrounding rock which can progressively crack, break, splinter and detach rock particles. This is followed by the decompression of the air as the wave retreats which can occur suddenly with explosive force which additionally weakens the rock. Cracks are gradually widened so each wave compresses more air, increasing the explosive force of its release. Thus, the effect intensifies in a 'positive feedback' system. Over time, as the cracks may grow they sometimes form a sea cave. The broken pieces that fall off produce two additional types of erosion, abrasion (sandpapering) and attrition. In corrasion, the newly formed chunks are thrown against the rock face. Attrition is a similar effect caused by eroded particles after they fall to the sea bed where they are subjected to further wave action. In coastal areas wave hydraulic action is often the most important form of erosion.

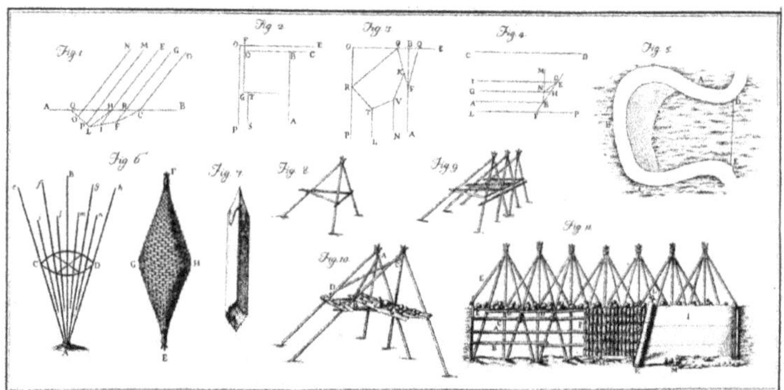

Tools to stem the erosion of rivers in the 18th century

Similarly, where hydraulic action is strong enough to loosen sediment along a stream bed and its banks; this will take rocks and particles from the banks and bed of the stream and add this to the stream's load. This process is the result of friction between the moving water and the static stream bed and banks. This friction increases with the speed of the water and the roughness of the bed. Once loosened the smaller particles are actually held in suspension by the force of the flowing water, these suspended particles can scour the sides and bottom of the stream. The scouring action produces distinctive markings on streams beds such as ripple marks, fluting, and crescent marks. The larger particles and even large rocks are *scooted* (dragged) along the bottom in a process known as *traction* which causes attrition, and are often "bounced" along in a process known as saltation where the force of the water temporarily lifts the rock particle which then crashes back into the bed dislodging other particles.

Hydraulic action also occurs as a stream tumbles over a waterfall to crash onto the rocks below. It usually leads to the formation of a plunge pool below the waterfall due in part to corrosion from the stream's load, but more to a scouring action as vortices form in the water as it escapes downstream. Hydraulic action can also cause the breakdown of river banks since there are water bubbles which enter the banks and collapse them when they expand.

Abrasion

Abrasion is the mechanical scraping of a rock surface by friction between rocks and moving particles during their transport by wind, glacier, waves, gravity, running water or erosion. After friction, the moving particles dislodge loose and weak debris from the side of the rock.

The intensity of abrasion depends on the hardness, concentration, velocity and mass of the moving particles.

Abrasion in Coastal Erosion

Coastal abrasion occurs as breaking waves which contain sand and larger fragments erode the shoreline or headland. This removes material resulting in undercutting and possible collapse of unsupported overhanging cliffs.

Abrasion by a Glacier

A glacier can "carve" a valley, cirque, or a tarn (glacial lake), by eroding rocks and soil and plucking them up. The rocks the glacier collects by plucking is then used as a tool to scrape even more debris from its environment. These glacial processes are very significant to the landscaping and erosion of earth, especially during the glacial periods.

Abrasion Platform

Abrasion platforms are shore platforms where wave action abrasion is a prominent process. If it is currently being fashioned, it will be exposed only at low tide, but there is a possibility that the wave-cut platform will be hidden sporadically by a mantle of beach shingle (the abrading agent). If the platform is permanently exposed above the high-water mark, it is probably a raised beach platform, which is not considered a product of abrasion.

Abrasion in Channel Transport

Abrasion in a stream or river channel occurs when the sediment carried by a river scours the bed and banks, contributing significantly to erosion. In addition to chemical weathering and the physical weathering of hydraulic action, freeze-thaw cycles, and more, there is a suite of processes which have long been considered to contribute significantly to bedrock channel erosion include plucking, abrasion (due to both bedload and suspended load), solution, and cavitation.

Bedload transport consists of mostly larger clasts, which cannot be picked up by the velocity of the stream flow, rolling, sliding, and/or saltating (bouncing) downstream along the bed. Suspended load typically refers to smaller particles, such as silt, clay, and finer grain sands uplifted by processes of sediment transport. Grains of various sizes and composition are transported differently in terms of the threshold flow velocities required to dislodge and deposit them, as is modeled in the Hjulström curve. These grains polish and scour the bedrock and banks when they make abrasive contact.

Abrasion from Wind

Much consideration has been given to the role of wind as an agent of geomorphological change on Earth and other planets. Aeolian processes involve wind eroding materials, such as exposed rock, and moving particles through the air to contact other materials and deposit them elsewhere. Mathematical models of these forces are notably similar to models in fluvial environments. Aeolian processes demonstrate their most notable consequences in arid regions of sparse vegetation and abundant unconsolidated sediments, such as sand. There is now evidence that bedrock canyons, landforms traditionally thought to evolve only from the fluvial forces of flowing water, may indeed be extended by the aeolian forces of wind, perhaps even amplifying bedrock canyon incision rates by an order of magnitude above fluvial abrasion rates. Redistribution of materials by wind occurs at multiple geographic scales and can have important consequences for regional ecology and landscape evolution.

Attrition (Erosion)

Attrition is a type of erosion that occurs when bed load erodes. It is characterized by wearing of land as well as the removal of dune sediments, rocks and other particles by the action of tidal currents, wave currents, high winds and other factors.

In attrition, rocks and other particles are carried downstream throughout river beds. This kind of movement causes these particles to be broken into tinier fragments. This phenomenon can also take place in glaciated regions, where attrition is typically caused by ice movement.

Attrition is a destructive process that wears away coastal regions. In this process, waves and other activities cause pebbles and rocks to bump against each other, resulting in fragmentation.

Attrition can produce both non-dramatic and dramatic erosion in regions where coastlines have fracture zones or rock layers with unstable erosion resistance. Typically, softer regions erode faster than harder areas. This leads to the formation of pillars, columns, bridges and tunnels. In some cases, abrasion also occurs in regions that have loose sand, strong winds and soft rocks.

The capacity of waves, currents and other variables to cause attrition depends on different factors. For instance, attrition can be controlled by the strength of rock and the existence of fractures and fissures. Other factors include fall debris rates that also depend on wave strength.

To control attrition, various measures can be employed such as the construction of seawalls. Other strategies include nourishment of coastal areas and using sandbags.

Internal Erosion

Internal erosion is the gradual diminution of the internal surface of a pipe due to the impingement of flowing particles. It is entirely a mechanical process, as opposed to corrosion and erosion-corrosion, which include chemical processes. Defects in underground pipes due to internal erosion may be corrected using trenchless rehabilitation techniques.

The term internal erosion is also commonly used to describe the removal of material by seepage from within an earthen dam.

The breakdown of pipe surfaces, both internal and external, are intensely studied by academics and engineers in fields where underground pipes are prevalent. Hydrocarbons make oil and gas pipelines especially susceptible to erosion and corrosion. Internal erosion can damage any kind of pipe, however, trenchless rehabilitation such as sliplining or mechanical spot repair can be used to address its effects.

Internal erosion is the primary cause of channeling in a pipe when the bottom of the pipe wears away, leaving behind a lengthwise gap along the pipe. Trenchless rehabilitation techniques such as sliplining can be used to insert a liner into the broken pipe.

Internal erosion can dramatically shorten the life of a pipe. Experts can make predictions on the time it might take to degrade a pipe based on the density and flow velocity of the fluid, along with other factors.

Basic Mechanisms of Internal Erosion

A consensual view in the community of water retaining structures is that four different basic processes can be identified within the general definition of internal erosion. These mechanisms are: backward erosion; concentrated leak erosion; suffusion; contact erosion. Their general definitions are the following.

- Backward erosion: Detachment of soil particles when the seepage exits to an unfiltered surface and leading to retrogressively growing pipes and sand boils;
- Concentrated leak erosion: Detachment of soil particles through a pre-existing path in the embankment or foundation;
- Suffusion: Selective erosion of the fine particles from the matrix of coarse particles;
- Contact erosion: Selective erosion of the fine particles from the contact with a coarser layer.

General conditions for occurrence of internal erosion

Two conditions should be fulfilled for internal erosion to occur.

- Fundamentally, the first condition is that particles can be detached, i.e. that hydraulic shear stresses are larger than resistant contact forces. To reach this hydro-mechanical criterion, water seeping through the flood defence must have sufficient velocity to provide the energy needed to detach particles from the soil structure.
- The second condition is that detached particles can be transported through the soil. Two criteria should be fulfilled. First, a hydro-mechanical criterion: flow is sufficient to carry the eroded particles. Second, a geometric criterion (which is specific to internal erosion): voids exist in the soils within the flood defence that are large enough for detached particles to pass through them. This void is either a pipe inside the soil, as in backward erosion or concentrated leak erosion, or pore space within the grains of a coarse layer, as observed in suffusion and contact erosion.

The nature of the soil in the embankment determines its vulnerability to erosion. Two main classes have to be distinguished:

- Granular non-cohesive soils: erosion resistance is related to particle buoyant weight and friction; hydro-mechanical transport criterion is linked to rolling and sliding resistance of the grains.

- Cohesive soils: erosion resistance is mainly related to attractive contact forces in between soil's particles; the main transport mode is suspension flow.

Successive Phases of Internal Erosion

The process of internal erosion of embankment dams or levees and their foundations can be represented by four phases. This applies for all types of internal erosion.

- Initiation: first phase of internal erosion, when one of the phenomena of detachment of particles occurs.
- Continuation: phase where the relationship of the particle size distribution between the base (core) material and the filter controls whether or not erosion will continue.
- Progression: phase of internal erosion, where hydraulic shear stresses within the eroding soil may or may not lead to the erosion process being on-going and in case of backward and concentrated leak erosion to formation of a pipe. The main issues are whether the pipe will collapse, or whether upstream zones may control the erosion process by flow limitation.
- Breach: final phase of internal erosion. It may occur by one of the following five phenomena (listed below in order of their observed frequency of occurrence).
- Gross enlargement of the pipe (which may include the development of a sinkhole from the pipe to the crest of the embankment).
- Slope instability of the downstream slope.
- Static liquefaction (which may include increase of pore pressure and sudden collapse in eroded zone).
- Unraveling of the downstream face.
- Overtopping (e.g. due to settlement of the crest from suffusion and/or due to the formation of a sinkhole from a pipe in the embankment).

Soil Erosion

Soil erosion is a process that involves the wearing away of the topsoil. The process involves the loosening of the soil particles, blowing or washing away of the soil particles, and either ends up in the valley and faraway lands or washed away to the oceans by rivers and streams. Soil erosion is a natural process which has increasingly been exacerbated by human activities such as agriculture and deforestation.

The wearing away of the topsoil is driven by erosion agents including the natural physical forces of wind and water, each contributing a substantial quantity of soil loss annually. Farming activities such as tillage also significantly contribute to soil erosion.

Thus, soil erosion is a continuous process and may occur either at a relatively unnoticed rate or an alarming rate contributing to copious loss of the topsoil. The outcomes of soil erosion are reduced agricultural productivity, ecological collapse, soil degradation, and the possibility of desertification.

Causes of Soil Erosion

All soils undergo soil erosion, but some are more vulnerable than others due to human activities and other natural causal factors. The severity of soil erosion is also dependent on the soil type and the presence of vegetation cover. Here are few of the major causes of soil erosion.

Rainfall and Flooding

Greater duration and intensity of rainstorm means greater potential for soil erosion. Rainstorm produces four major types of soil erosion including rill erosion, gully erosion, sheet erosion, and splash erosion. These types of erosions are caused by the impacts of raindrops on the soil surface that break down and disperse the soil particles, which are then washed away by the storm water runoff.

Over time, repeated rainfall can lead to significant amounts of soil loss. Rapidly moving storm water, flashfloods, and flooding may also occur because of excess surface water runoff, thus, causing extreme local erosion by plucking bed rocks, forming rock cut-basins, creating potholes, and washing away the loosened soil particles.

Rivers and Streams

The flow of rivers and streams causes valley erosion. The water flowing in the rivers and streams tend to eat away the soils along the water systems leading to a V-shaped erosive activity. When the rivers and streams are full of soil deposits due to sedimentation and the valley levels up with the surface, the water ways begin to wash away the soils at the banks.

This erosive activity is termed as lateral erosion which extends the valley floor and brings about a narrow floodplain. This erosive activity is evident in most rivers or streams especially during heavy rainfall and rapid river channel movement.

High Winds

High winds can contribute to soil erosion, particularly in dry weather periods or in the arid and semi-arid (ASAL) regions. The wind picks up the loose soil particles with its natural force and carries them away to far lands, leaving the soil sculptured and denudated. It is severe during the times of drought in the ASAL regions. Hence, wind erosion is a major source of soil degradation and desertification.

Overgrazing, Overstocking and Tillage Practices

The transformation of natural ecosystems to pasture lands has largely contributed to increased rates of soil erosion and the loss of soil nutrients and the top soil. Overstocking and overgrazing has led to reduced ground cover and break down of the soil particles, giving room for erosion and accelerating the erosive effects by wind and rain. This reduces soil quality and agricultural productivity.

Agricultural tillage depending on the machinery used also breaks down the soil particles, making the soils vulnerable to erosion by water. Up and down field tillage practices as well create pathways for surface water runoff and can speed up the soil erosion process.

Deforestation, Reduced Vegetation Cover, and Urbanization

Deforestation and urbanization destroy the vegetation land cover. Agricultural practices such as burning and clearing of vegetation also reduce the overall vegetation cover. As a result, the lack of land cover causes increased rates of soil erosion.

Trees and vegetation cover help to hold the soil particles together thereby reduces the erosive effects of erosion caused by rainfall and flooding. Deforestation and urbanization are some of the human actions that have continued the cycle of soil loss.

Mass Movements and Soil Structure/Composition

The outward and downward movements of sediments and rocks on slanting or slope surfaces due to gravitational pull qualify as an important aspect of the erosion process. This is because mass movements aids in the breakdown of the soil particles that makes them venerable to water and wind erosion. Soil structure and composition is another factor that determines erosivity of wind or rainfall.

For instance, clay soils tend to be more resistant to soil erosion compared to sandy or loose silt soils. Soil moisture content and organic matter make up are some of the soil component aspects that determine erosivity of wind or rainfall.

Effects of Soil Erosion

The consequences of soil erosion are primarily centered on reduced agricultural productivity as well as soil quality. Water ways may also be blocked, and it may affect water quality. This means most of the environmental problems the world face today arises from soil erosion. The effects of soil erosion include:

Loss of Arable Land

Lands used for crop production have been substantially affected by soil erosion. Soil erosion eats away the top soil which is the fertile layer of the land and also the component that supports the soil's essential microorganisms and organic matter. In this view, soil erosion has severely threatened the productivity of fertile cropping areas as they are continually degraded.

Because of soil erosion, most of the soil characteristics that support agriculture have been lost, causing ecological collapse and mass starvation. It is likely that most of the cultivated areas around the globe are vulnerable to soil erosion.

Water Pollution and Clogging of Waterways

Soils eroded from agricultural lands carry pesticides, heavy metals, and fertilizers which are washed into streams and major water ways. This leads to water pollution and damage to marine and freshwater habitats. Accumulated sediments can also cause clogging of water ways and raises the water level leading to flooding.

The water quality of various streams, rivers, and coastal areas has also been deteriorated as a result of soil erosion, eventually affecting the health of the local communities.

Sedimentation and Threat to Aquatic Systems

Apart from polluting the water systems, high soil sedimentation can be catastrophic to the survival of aquatic life forms. Silt can smother the breeding grounds of fish and equally lessens their food supply since the siltation reduces the biodiversity of algal life and beneficial aquatic plants. Sediments may also enter the fish gills, affecting their respiratory functions.

Air Pollution

Wind erosion picks up dust particles of the soil and throws them into the air, causing air pollution. Some of the dust particles may contain harmful and toxic particles such as petroleum and pesticides that can pose a severe health hazard when inhaled or ingested.

Dust plumes from the deserts or dry areas can cause large and widespread air pollution as the winds move. Such a case is evident in North America where dust winds from the Gobi desert have recurrently been a serious problem.

Destruction of Infrastructure

Soil erosion can affect infrastructural projects such as dams, drainages, and embankments. The accumulation of soil sediments in dams/drainages and along embankments can reduce their operational lifetime and efficiency. Also, the silt up can support plant life that can, in turn, cause cracks and weaken the structures. Soil erosion from surface water runoff often causes serious damage to roads and tracks, especially if stabilizing techniques are not used.

Desertification

Soil erosion is a major driver of desertification. It gradually transforms a habitable land and the ASAL regions into deserts. The transformations are worsened by the destructive use of the land and deforestation that leaves the soil naked and open to erosion. This usually leads to loss of biodiversity, alteration of ecosystems, land degradation, and huge economic losses.

Sheet Erosion

Sheet erosion is the detachment of soil particles by raindrop impact and their removal downslope by water flowing overland as a sheet instead of in definite channels or rills.

A more or less uniform layer of fine particles is removed from the entire surface of an area, sometimes resulting in an extensive loss of rich topsoil. Sheet erosion commonly occurs on recently plowed fields or on other sites having poorly consolidated soil material with scant vegetative cover.

There are two stages of sheet erosion. The first is rain splash, in which soil particles are knocked into the air by raindrop impact. A hundred tons of particles per acre may be dislodged during a single rainstorm. In the second stage, the loose particles are moved downslope, commonly by sheet flooding. Broad sheets of rapidly flowing water filled with sediment present a potentially high erosive force. Generally produced by cloudbursts, sheet floods are of brief duration, and they commonly move only short distances. On relatively rough surfaces, sheet flooding may give way to rill wash, in which the water moves in a system of enmeshed micro channels, which eventually become larger and develop into gullies.

Sheet erosion is the uniform removal of soil in thin layers by the forces of raindrops and overland flow. It can be a very effective erosive process because it can cover large areas of sloping land and go unnoticed or insidious for quite some time or because it is it is difficult to see or measure. Farmers are seldom aware of soil loss until farm fields begin to change color or subsoil becomes mixed with the topsoil.

Sheet erosion can be recognized by either soil deposition at the bottom of a slope, or by the presence of light - colored subsoil appearing on the surface.

If left unattended, sheet erosion will gradually remove the nutrients and organic matter, which are important to agriculture and eventually lead to unproductive soil (figure below).

Figure. Sheet erosion removes a fairly uniform layer (a sheet) of soil from the land surface by raindrop splashes or MP3- Measurements of sheet erosion

Causes and Dynamics of Sheet Erosion

Sheet erosion is caused by the force of raindrops impacting on bare soil and dislodging particles of earth. This force is dependent on the speed of fall (a function of the length of fall and the windspeed) and the weight (a function of the diameter of drops). After falling for 10 meters raindrops reach 90% of their final speed, which is determined by the balance between gravity and the air resistance of the bearing surface of the drop.

The resistance of the soil material will depend on the presence of pebbles, the percentage of silt and fine sand (2-100 μ), organic matter and clay, the presence of gypsum or limestone, iron hydroxides and free aluminum, and again the structural stability and permeability of the soil profile.

Particles are initially carried a short distance by the splash effect and then by sheet runoff. The impact of raindrops sends droplets and particles in all directions, but on slopes the distance covered uphill is less than that downhill, so that on the whole particles move downhill in jumps.

Process of Sheet Erosion

Raindrop action on bare soil disrupts aggregates, dislodges soil particles and compact the erodible soil surface. If rainfall exceeds infiltration, a surface film of water forms, building up into flows 2-3 mm deep. Continuing rainfall causes turbulence within the flow that may increase the water's erosive effect up to 200 times.

Sheet erosion will continue and the removal of a uniform thin layer of soil by raindrop splash or water run-off will increase. This thin layer of topsoil often disappears gradually, making it difficult to monitor because the damage is not immediately perceptible. This insidious process is often overlooked until the subsoil is exposed.

Impacts of Sheet Erosion

Loss of the finest soil particles, to which the bulk of plant-available nutrients and organic matter adhere, affects the productivity of the land.

Erosion may also result in removal of seeds or seedlings and reduce the soil's ability to store water for plants to draw upon between rainfall events.

Soil deposited off-site through this type of erosion causes crop and pasture damage, water-quality deterioration and stream, dam, lake and reservoir sedimentation.

As organic matter is removed and vegetation cover decreases, aggregates break down and many soils begin to form a crust, which air and water can no longer penetrate. Between grass tufts, small but increasing crusted areas will appear and in sloping areas this often begins at mid slope. Other clues include sorting of soil particles by water leaving small heaps of washed sand, and light clay-particle coating in small depressions.

Headward Erosion

Headward Erosion is a process of erosion that lengthens a stream, a valley or a gully at its beginning and also enlarges its drainage basin.

When water falls on a planer surface, it has a tendency to flow along slope in form of an oscillating stream. In place of origin of stream formation (from sheet flow) the water (by solution, abrasion, collision etc.) begins to erode some part of the underlying rock.

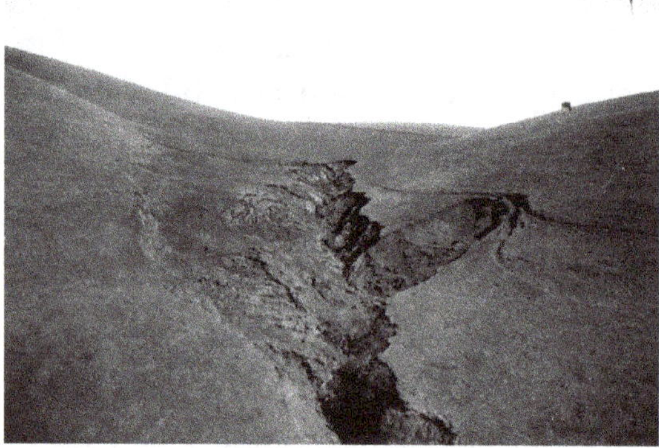

This erosion creates micro-depressions the slope of these micro-depressions have steeper slopes compared with surrounding planer surface. Such slopes further aid in the erosion along the river source.

When the consequent sheet wash drains the planer surface, the process is repeated. Intuitively we assume that a stream erodes downstream channel.

Stream Capture- The Development of a Watergap

Stream Capture is a geomorphic phenomenon that occurs when a stream or river from a neighboring drainage system or watershed erodes through the divide between two streams and "captures" another stream which then is diverted from its former bed and now flows down the bed of the capturing stream. This can occur through headward erosion of one stream valley into another or through lateral erosion of meanders through the sediments between parallel streams.

An example: -

Stage 1: Beaverdam Creek, Gap Run and Goose Creek flow eastward through the Blue Ridge and enter the Potomac.

Stage 2: As the land is eroded downward, the three east flowing creeks do not have the power to erode as far through the Blue Ridge as the Shenandoah, Potomac system. The Shenandoah extends itself southward by headward erosion through the relatively high land west of the Blue Ridge. It eventually captures Beaverdam creek.

Stage 3: The capture of Beaverdam creek added more discharge to the Shenandoah which was able to therefore erode more. Headward erosion leads to the capture of Gap Run. The water gaps where Beaverdam creek and Gap Run used to flow through the Blue Ridge are left as wind gaps.

Stage 4: Eventually Goose Creek is captured as well Snicker's Gap, Ashby Gap and Manassas Gap are left as wind gaps. As the land on either side of the ridge is eroded down together with the ridge summit, the relative elevation of the wind gaps becomes higher and higher.

Stream Types Created by Headward Erosion

Three kinds of streams are formed by headward erosion.

1. Insequent streams form by random headward erosion, usually from sheet flow of water over the landform surface. The water collects in channels where the velocity and erosional power increase, cutting into and extending the heads of gullies.
2. Subsequent streams form by selective headward erosion by cutting away at less resistive rocks in the terrain.
3. Obsequent and resequent streams form after time in an area of insequent or subsequent streams. Obsequent streams are insequent streams that now flow in an opposite direction of the original drainage pattern. Resequent streams are subsequent streams that have also changed direction from their original drainage patterns.

Drainage Patterns Created by Headward Erosion

Headward erosion creates three major kinds of drainage patterns: dendritic patterns, trellis patterns, and rectangular and angular patterns.

1. Dendritic patterns form in homogenous landforms where the underlying bedrock has no structural control over where the water flows. They have a very characteristic pattern of branching at acute angles with no common or similarly repeating pattern.
2. Trellis patterns form in where the underlying bedrock where there is repeating weaker and stronger types of rock. The trellis pattern cuts down deeper into the weaker bedrock, and is characterized by nearly parallel streams that branch at higher angles.
3. Rectangular and angular patterns are characterized by branching of tributaries at nearly right angles and tributaries which themselves exhibit right-angle bends in their channels. These usually form in jointed igneous bedrocks, horizontal sedimentary beds with well-developed jointing or intersecting faults.

Four minor kinds of drainage patterns also can be created: radial patterns, annular patterns, centripetal patterns and parallel patterns.

1. Radial patterns are characterized by flow of water outward from a central point, such as down a newly formed cinder volcano cone or an intrusive dome.

2. Annular patterns form on domes of alternating weak and hard bedrocks. The pattern formed is similar to that of a bulls eye when viewed from above, as the weaker bedrocks are eroded and the harder are left in place.
3. Centripetal patterns form where water flows into a central location, such as in a karst limestone terrain where the water flows down into a sinkhole and then underground.
4. Parallel patterns are not very common and form on unidirectional regional slope or parallel landform features. They are usually limited to a small, generalized area.

Glacial Erosion

Glaciers are effective agents of erosion, especially in situations where the ice is not frozen to its base and can therefore slide over the bedrock or other sediment. The ice itself is not particularly effective at erosion because it is relatively soft; instead, it is the rock fragments embedded in the ice and pushed down onto the underlying surfaces that do most of the erosion. A useful analogy would be to compare the effect of a piece of paper being rubbed against a wooden surface, as opposed to a piece of sandpaper that has embedded angular fragments of garnet.

The results of glacial erosion are different in areas with continental glaciation versus alpine glaciation. Continental glaciation tends to produce relatively flat bedrock surfaces, especially where the rock beneath is uniform in strength. In areas where there are differences in the strength of rocks, a glacier obviously tends to erode the softer and weaker rock more effectively than the harder and stronger rock. Much of central and eastern Canada, which was completely covered by the huge Lauren tide Ice Sheet at various times during the Pleistocene, has been eroded to a relatively flat surface. In many cases the existing relief is due the presence of glacial deposits — such as drumlins, eskers, and moraines — rather than to differential erosion (figure below).

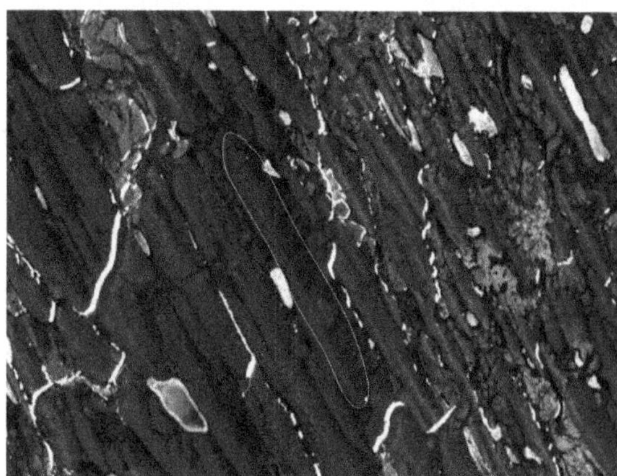

Figure: Drumlins — streamlined hills formed beneath a glacier, here made up of sediment — in the Amundsun Gulf region of Nunavut. The drumlins are tens of meters high, a few hundred meters across, and a few kilometers long. One of them is highlighted with a dashed white line.

Alpine glaciers produce very different topography than continental glaciers, and much of the topographic variability of western Canada can be attributed to glacial erosion. In general, glaciers are much wider than rivers of similar length, and since they tend to erode more at their bases than their sides, they produce wide valleys with relatively flat bottoms and steep sides — known as U-shaped valleys (figure below). Howe Sound, north of Vancouver, was occupied by a large glacier that originated in the Squamish, Whistler, and Pemberton areas, and then joined the much larger glacier in the Strait of Georgia. Howe Sound and most of its tributary valleys have pronounced U-shaped profiles.

Figure: A depiction of a U-shaped valley occupied by a large glacier.

Glaciers only move very slowly but are still capable of tremendous amount of erosion. There are three main types of glacial erosion - plucking, abrasion and freeze thaw.

Plucking

As the glacier moves, friction causes the bottom of the glacier to melt this water freezes into joints in the rock. When the glacier moves again the rock is pulled away or 'plucked' from the base of the valley.

Abrasion

Rock pieces that have been plucked away and carried by the glacier (moraine) act like sandpaper scraping away at the valley bottom and sides.

Freeze – thaw or frost shattering

Water enters into cracks in the rock over time it freezes and repeatedly thaws putting large amounts of pressure on the rock – eventually this will force the rock to break apart or shatter.

Deposition

Deposition is the geological process in which sediments, soil and rocks are added to a landform or land mass. Previously eroded sediment will be transported by wind, ice, water which loses its kinetic energy in fluid and thus deposited. Geologic deposition includes Beach sand, lake mud, sand dunes, glacial moraines, river deltas, gravel bars, and coal deposits.

The Elements of Deposition

Several elements affect when and where deposition occurs once rocks erode. The velocity, or speed of wind and water plays an important part because as they slow, heavier sediments drop out and are deposited. The thickness, heaviness and size of sediment also affect the rate of deposition. Larger and denser particles are heavier and land first before, less dense particles. The shape of sediment also affects deposition rates, as round pieces of sediment settle more quickly than flat pieces.

Landforms Produced by Deposition

Deposition creates many types of landforms on earth. Gravity and weight create rockslides on the sides of hills and mountains, depositing rocks at the bottom. Wind's efforts create patterns in the sand dunes of the desert as it moves the sand across the surface. Rivers create deltas when they deposit sand and sediment at their mouths, where the water slows to meet the ocean. Ocean waves create beaches and sand bars as they deposit sand over time.

Examples of Deposition of Sediments by Wind and Water

Sand Dunes

The sand blown by wind deposits in the form of a hill or ridge. They are longer in length on the side of the wind. The depression between two dunes is called a slack.

- Guadalupe-Nipomo Dunes located in San Francisco, United States.
- Silver lake sand dunes located in Michigan, United States

Sand Ripples

They are formed by the action of wind or water and are shorter in length than sand dunes. They are either symmetrical or asymmetrical in nature.

- Crescent-shaped rippled Barchans dunes in Afghanistan
- Ripple formation on the North Carolina beach, Carolina

Sand Drift

The sand gets drifted due to wind or water. This results in formations having varying shapes and sizes. The typical shapes include hummock, sand pile, or a knoll. They are smaller than sand dunes.

- Drift at New Brighton in Canterbury, New Zealand

Beach

A landform formed along the coast of a water body such as an ocean is called a beach. The sand contained by the beach is deposited after the erosion of rocks and coral reefs.

- The long stretch of Paradise Island beach, Bahamas
- Mar del Plata beach, Buenos Aires , Argentina

Loess

- The wind deposits sediments that possess a yellowish-gray color. These deposits are known to be having a high thickness and they create a wall-like structure called loess.
- A loess structure formed in Vicksburg, Mississippi, United States
- A loess plateau formed at Hunyuan, Shanxi Province, China

Floodplain

It is the area contiguous to a flowing water body such as a river, which receives plenty of water during high water levels.

- Paraná Delta is a floodplain in succession to the Paraná River, Brazil
- Maputo Special Reserve area in Mozambique, Africa

Levees

It is a landform that prevents water bodies from causing floods. It is naturally occurring or sometimes built by man-made processes to protect certain portions of land from floods.

- A strip of dry land covered by the levee formed at Yangtze River, China
- Levee at Sacramento (the city known for catastrophic flooding), Canada

Glacier

At times, a water body flowing in cold regions freezes into ice. This is called a glacier. It contains a huge ice mass, which includes sediments as well.

- Timpanogos Glacier located at Wasatch range of Utah, United States
- Large glacier named Perito Moreno located in Western Patagonia, Argentina

Alluvial Fan

It is a fan-shaped structure formed when water flows through hills or mountains. It is known to comprise sand, gravel, or other types of sediments.

- An alluvial fan of a large size at Tibetan plateau, China
- The alluvial fan formation found near Lake Louise, Alberta, Canada

Delta

This is a structure formed at the mouth of the water bodies such as a river. It is a result of deposition of alluvial sediments before the water plunges into a larger water body such as an ocean.

- Delta formed by the Nile River while flowing to the Mediterranean Sea
- Mississippi Delta formed by the river, Mississippi, United States

Null-point Hypothesis

The null-point hypothesis explains how sediment is deposited throughout a shore profile according to its grain size. This is due to the influence of hydraulic energy, resulting in a seaward fining of sediment particle size, or where fluid forcing equals gravity for each grain size. The concept can also be explained as "sediment of a particular size may move across the profile to a position where it is in equilibrium with the wave and flows acting on that sediment grain". This sorting mechanism combines the influence of the down-slope gravitational force of the profile and forces due to flow asymmetry; the position where there is zero net transport is known as the null point and was first proposed by Cornaglia in 1889. Figure Below.

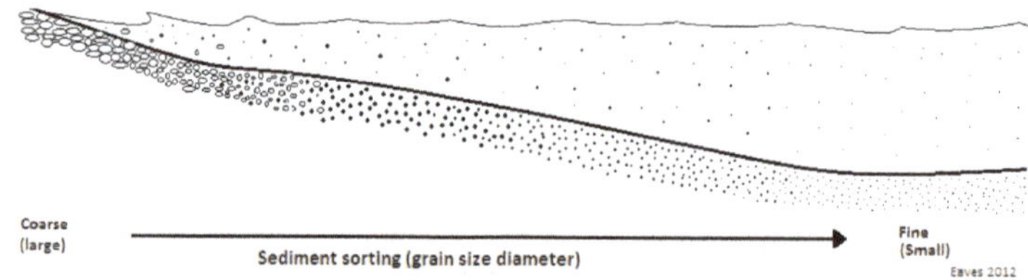

Figure. Illustrating the sediment size distribution over a shoreline profile, where finer sediments are transported away from high energy environments and settle out of suspension, or deposit in calmer environments. Coarse sediments are maintained in the upper shoreline profile and are sorted by the wave-generated hydraulic regime.

The first principle underlying the null point theory is due to the gravitational force; finer sediments remain in the water column for longer durations allowing transportation outside the surf zone to deposit under calmer conditions. The gravitational effect, or settling velocity determines the location of deposition for finer sediments, whereas a grain's internal angle of friction determines the deposition of larger grains on a shore profile. The secondary principle to the creation of seaward sediment fining is known as the hypothesis of asymmetrical thresholds under waves; this describes the interaction between the oscillatory flow of waves and tides flowing over the wave ripple bedforms in an asymmetric pattern. "The relatively strong onshore stroke of the wave forms an eddy or vortex on the lee side of the ripple, provided the onshore flow persists, this eddy remains trapped in the lee of the ripple. When the flow reverses, the eddy is thrown upwards off the bottom and a small cloud of suspended sediment generated by the eddy is ejected into the water column above the ripple, the sediment cloud is then moved seaward by the offshore stroke of the wave." Where there is symmetry in ripple shape the vortex is neutralized, the eddy and its associated sediment cloud develops on both sides of the ripple. This creates a cloudy water column which travels under tidal influence as the wave orbital motion is in equilibrium.

The Null-point hypothesis has been quantitatively proven in Akaroa Harbour, New Zealand, The Wash, U.K., Bohai Bay and West Huang Sera, Mainland China, and in numerous other studies; Ippen and Eagleson, Eagleson and Dean and Miller and Zeigler.

Deposition of Non-cohesive Sediments

Large-grain sediments transported by either bed load or suspended load will come to rest when there is insufficient bed shear stress and fluid turbulence to keep the sediment moving; with the suspended load this can be some distance as the particles need to fall through the water column. This is determined by the grain's downward acting weight force being matched by a combined buoyancy and fluid drag force and can be expressed by:

$$\frac{4}{3}\pi R^3 \rho_s g = \frac{4}{3}\pi R^3 \rho g + \frac{1}{2}\mathbb{C}_d \rho \pi R^2 w_s^2$$

Downward acting weight force = Upward-acting buoyancy force + Upward-acting fluid drag force

Where:

- π is the ratio of a circle's circumference to its diameter.
- R is the radius of the spherical object (in m),
- ρ is the mass density of the fluid (kg/m^3),
- g is the gravitational acceleration (m/s^2),
- C_d is the drag coefficient, and
- w_s is the particle's settling velocity (in m/s).

In order to calculate the drag coefficient, the grain's Reynolds number needs to be discovered, which is based on the type of fluid through which the sediment particle is flowing, laminar flow, turbulent flow or a hybrid of both. When the fluid becomes more viscous due to smaller grain sizes or larger settling velocities, prediction is less straightforward and it is applicable to incorporate Stokes Law (also known as the frictional force, or drag force) of settling.

Deposition of Cohesive Sediments

Cohesion of sediment occurs with the small grain sizes associated with silts and clays, or particles smaller than 4ϕ on the phi scale. If these fine particles remain dispersed in the water column, Stokes law applies to the settling velocity of the individual grains, although due to sea water being a strong electrolyte bonding agent, flocculation occurs where individual particles create an electrical bond adhering each other together to form flocs. "The face of a clay platelet has a slight negative charge where the edge has a slight positive charge, when two platelets come into close proximity with each other the face of one particle and the edge of the other are electrostatically attracted."- Flocs then have a higher combined mass which leads to quicker deposition through a higher fall velocity, and deposition in a more shoreward direction than they would have as the individual fine grains of clay or silt.

The Occurrence of Null Point Theory

Akaroa Harbour is located on Banks Peninsula, Canterbury, New Zealand. The formation of this harbour has occurred due to active erosional processes on an extinct shield volcano, whereby the sea has flooded the caldera, creating an inlet 16 km in length, with an average width of 2 km and a depth of −13 m relative to mean sea level at the 9 km point down the transect of the central axis. The predominant storm wave energy has unlimited fetch for the outer harbour from a southerly direction, with a calmer environment within the inner harbour, though localized harbour breezes create surface currents and chop influencing the marine sedimentation processes. Deposits of loess from subsequent glacial periods have in filled volcanic fissures over millennia, resulting in volcanic basalt and loess as the main sediment types available for deposition in Akaroa Harbour.

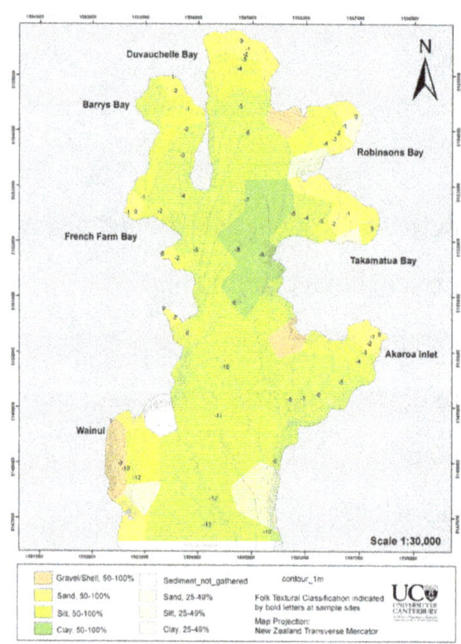

Figure. Map of Akaroa Harbour showing a fining of sediments with increased bathymetry toward the central axis of the harbour.

Hart discovered through bathymetric survey, sieve and pipette analysis of sub tidal sediments, that sediment textures were related to three main factors: depth, distance from shoreline, and distance along the central axis of the harbour. This resulted in the fining of sediment textures with increasing depth and towards the central axis of the harbour, or if classified into grain class sizes, "the plotted transect for the central axis goes from silty sands in the intertidal zone, to sandy silts in the inner near shore, to silts in the outer reaches of the bays to mud at depths of 6 m or more".

Other studies have shown this process of the winnowing of sediment grain size from the effect of hydrodynamic forcing; Wang, Collins and Zhu qualitatively correlated increasing intensity of fluid forcing with increasing grain size. "This correlation was demonstrated at the low energy clayey tidal flats of Bohai Bay (China), the moderate environment of the Jiangsu coast (China) where the bottom material is silty, and the sandy flats of the high energy coast of The Wash (U.K.)." This research shows conclusive evidence for the null point theory existing on tidal flats with differing hydrodynamic energy levels and also on flats that are both erosional and accretional.

Kirby R. takes this concept further explaining that the fines are suspended and reworked aerially offshore leaving behind lag deposits of mainly bivalve and gastropod shells separated out from the finer substrate beneath, waves and currents then heap these deposits to form chenier ridges throughout the tidal zone, which tend to be forced up the foreshore profile but also along the foreshore. Cheniers can be found at any level on the foreshore and predominantly characterize an erosion-dominated regime.

Applications for Coastal Planning and Management

The null point theory has been controversial in its acceptance into mainstream coastal science as the theory operates in dynamic equilibrium or unstable equilibrium, and many field and laboratory observations have failed to replicate the state of a null point at each grain size throughout the profile. The interaction of variables and processes over time within the environmental context causes issues; "the large number of variables, the complexity of the processes, and the difficulty in observation, all place serious obstacles in the way of systematization, therefore in certain narrow fields the basic physical theory may be sound and reliable but the gaps are large."

Geomorphologists, engineers, governments and planners should be aware of the processes and outcomes involved with the null point hypothesis when performing tasks such as beach nourishment, issuing building consents or building coastal defence structures. This is because sediment grain size analysis throughout a profile allows inference into the erosion or accretion rates possible if shore dynamics are modified. Planners and managers should also be aware that the coastal environment is dynamic and contextual science should be evaluated before implementation of any shore profile modification. Thus theoretical studies, laboratory experiments, numerical and hydraulic modelling seek to answer questions pertaining to littoral drift and sediment deposition, the results should not be viewed in isolation and a substantial body of purely qualitative observational data should supplement any planning or management decision.

Desert Varnish

Desert varnish is the thin red-to-black coating found on exposed rock surfaces in arid regions. Varnish is composed of clay minerals, oxides and hydroxides of manganese and/or iron, as well as other particles such as sand grains and trace elements. The most distinctive elements are manganese (Mn) and iron (Fe).

Bacteria take manganese out of the environment, oxidize it, and cement it onto rock surfaces. In the process, clay and other particles also become cemented onto the rock. These bacteria microorganisms live on most rock surfaces.

The sources for desert varnish components come from outside the rock, most likely from atmospheric dust and surface runoff. Streaks of black varnish often occur where water cascades over cliffs. No major varnish characteristics are caused by wind.

The color of rock varnish depends on the relative amounts of manganese and iron in it: manganese-rich varnishes are black; iron-rich varnishes are red or orange; varnishes with similar amounts of manganese and iron are some shade of brown. Varnish surfaces tend to be shiny when the varnish is smooth and rich in manganese.

Desert varnish often served as canvases for American Indians, who carved petroglyphs onto the shiny surfaces.

A complete coat of manganese-rich desert varnish takes thousands of years, so it is rarely found on easily eroded surfaces. A change to more acidic conditions (such as acid rain) can erode rock varnish. Lichens can also chemically erode rock varnish, as can visitors who scratch graffiti into it.

References

- Monroe, James Stewart, Reed Wicander, & Richard W. Hazlett. (2011) Physical Geology: Exploring the Earth. Cengage Learning ISBN 9781111795658. pg 465,591
- Causes-and-effects-of-soil-erosion, environment: eartheclipse.com, Retrieved 14 April 2018
- Raeside, J. D. (1964). "Loess Deposits of the South Island, New Zealand, and Soils Formed on them". New Zealand Journal of Geology and Geophysics. 7 (4): 811–838. doi:10.1080/00288306.1964.10428132. ISSN 0028-8306
- Measurements-of-Sheet-Erosion-Training-Manual-256803023: researchgate.net, Retrieved 26 June 2018
- Oldale, Robert N. (1999). "Coastal Erosion on Cape Cod: Some Questions and Answers". Cape Naturalist, the journal of the Cape Cod Museum of Natural History. U.S. Geological Survey. 25: 70–76. Retrieved 15 October 2016
- Landforms-glacial-erosion, geography: revisionworld.com, Retrieved 11 April 2018

- Jolliffe, I. P. (1978). "Littoral and offshore sediment transport". Progress in Physical Geography. 2 (2): 264–308. doi:10.1177/030913337800200204. ISSN 0309-1333
- Deposition-kids-8512606: sciencing.com, Retrieved 21 May 2018
- Bennett, Matthew M., & Neil F. Glasser. Glacial Geology: Ice Sheets and Landforms. (2011) Ch. 5 Glacial abrasion. John Wiley & Sons. ISBN 9781119966692
- Understanding-deposition-in-geology-with-examples: sciencestruck.com, Retrieved 19 July 2018
- Horn, Diane P (1992). "A review and experimental assessment of equilibrium grain size and the ideal wave-graded profile". Marine Geology. 108 (2): 161–174. doi:10.1016/0025-3227(92)90170-M. ISSN 0025-3227

Permissions

All chapters in this book are published with permission under the Creative Commons Attribution Share Alike License or equivalent. Every chapter published in this book has been scrutinized by our experts. Their significance has been extensively debated. The topics covered herein carry significant information for a comprehensive understanding. They may even be implemented as practical applications or may be referred to as a beginning point for further studies.

We would like to thank the editorial team for lending their expertise to make the book truly unique. They have played a crucial role in the development of this book. Without their invaluable contributions this book wouldn't have been possible. They have made vital efforts to compile up to date information on the varied aspects of this subject to make this book a valuable addition to the collection of many professionals and students.

This book was conceptualized with the vision of imparting up-to-date and integrated information in this field. To ensure the same, a matchless editorial board was set up. Every individual on the board went through rigorous rounds of assessment to prove their worth. After which they invested a large part of their time researching and compiling the most relevant data for our readers.

The editorial board has been involved in producing this book since its inception. They have spent rigorous hours researching and exploring the diverse topics which have resulted in the successful publishing of this book. They have passed on their knowledge of decades through this book. To expedite this challenging task, the publisher supported the team at every step. A small team of assistant editors was also appointed to further simplify the editing procedure and attain best results for the readers.

Apart from the editorial board, the designing team has also invested a significant amount of their time in understanding the subject and creating the most relevant covers. They scrutinized every image to scout for the most suitable representation of the subject and create an appropriate cover for the book.

The publishing team has been an ardent support to the editorial, designing and production team. Their endless efforts to recruit the best for this project, has resulted in the accomplishment of this book. They are a veteran in the field of academics and their pool of knowledge is as vast as their experience in printing. Their expertise and guidance has proved useful at every step. Their uncompromising quality standards have made this book an exceptional effort. Their encouragement from time to time has been an inspiration for everyone.

The publisher and the editorial board hope that this book will prove to be a valuable piece of knowledge for students, practitioners and scholars across the globe.

Index

R

S

T

V

W

www.ingramcontent.com/pod-product-compliance
Lightning Source LLC
Chambersburg PA
CBHW082035190326
41458CB00010B/3374

* 9 7 8 1 6 4 1 7 2 0 9 8 4 *